RELENTLESS ENEMIES

Lions and Buffalo

RELENTLESS ENEMIES

Lions and Buffalo

PHOTOGRAPHED BY BEVERLY JOUBERT AND WRITTEN BY DERECK JOUBERT

NATIONAL GEOGRAPHIC

Washington, D.C.

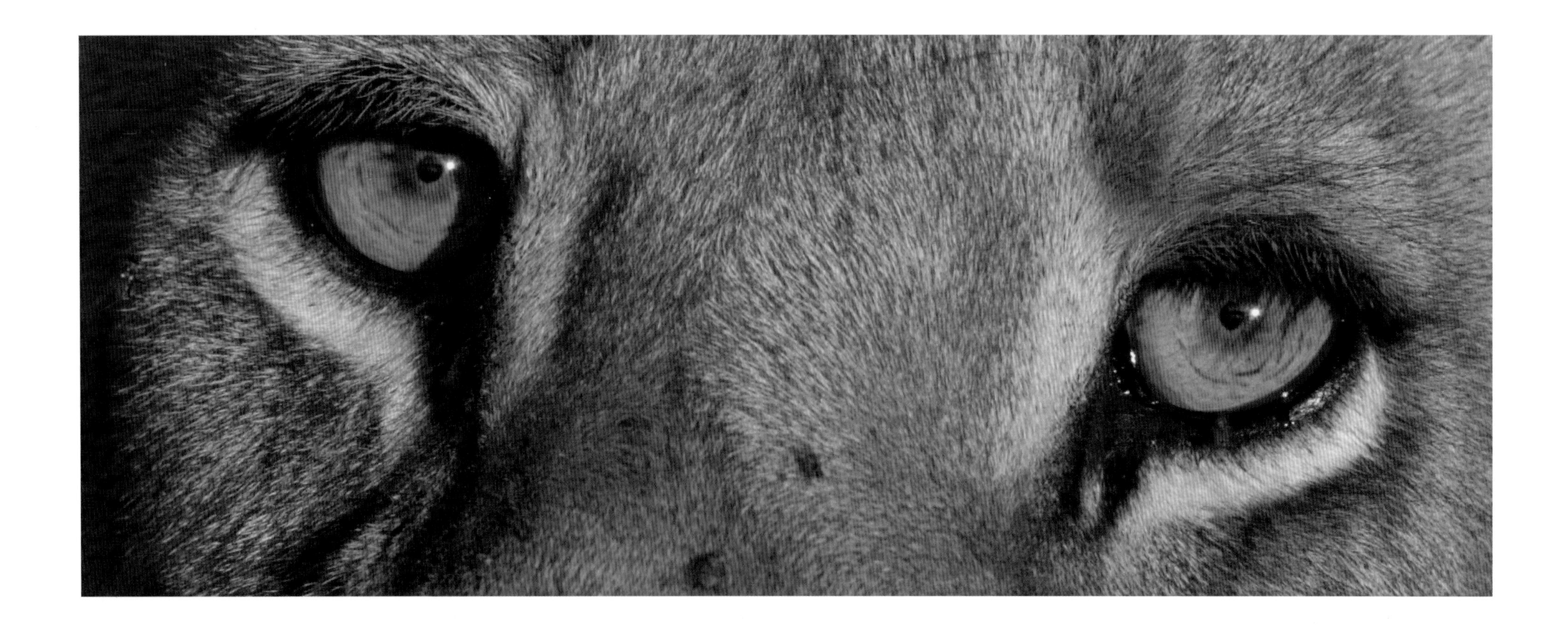

Page 1: *A lattice of grass is the best camouflage for a lion. Focused, and dissolving in and out of reality like a ghost of the savanna, she stalks her quarry.*

Page 2: *Duba Plains is laid out at its finest, half flooded under a layer of winter floodwater that the lions must wade through daily to get to the buffalo they need to hunt. They are a team, with a single goal.*

Page 6: *Alone against the hordes, a single female is ready at dawn for the hunt, ready to take on the biggest of the big.*

Duba at dawn: dust and mist mingle at the buffalo's feet in a unique place in a harsh and stimulating time.

At dawn, buffalo are tightly bunched together in golden dust. **OPPOSITE:** *A Duba lion is veiled in mist*

Foreword

SINCE I FIRST CAME ACROSS DERECK AND BEVERLY filming along the Linyanti River about 20 years ago, I have considered them to be pioneers in what they do: living and filming in the bush, immersing themselves fully in the experience, not because it feels good to them, but for all of us. They have always said that unless they could fully understand and experience the highs and the lows of a lifestyle like this, they would never be able to honestly portray it in their books and films. I think that this comes across in their work very strongly. From time to time we'd meet in different locations around Botswana and swap tales of our adventures, and when I traveled up to see them there was always something fascinating and new from them, a testament to the amount of time they physically spend with their subjects in the wild.

From our distant perspective wildlife and nature often seem to be one-dimensional, conforming to the rules of what science discovers about them. Only by spending extraordinary amounts of time working within nature can you see beyond the rules, and this is what the Jouberts do best. The wild places of the world are there as a reflection for us, something important, not to be abused, or simply for our entertainment. These precious jewels, of which Botswana has some of the best and most well looked after in the world, need to be protected for their own intrinsic value, not simply as a revenue source. The films and books of the Jouberts, like this one, do many things for us. They impart the preciousness of these jewels—places like Duba Plains—and lead us to a greater appreciation of the value of wild places. These projects make us think as much about ourselves as about the lions and buffalo or elephants that are so often the subject of their work.

We in Botswana believe in the careful protection of our resources. The low-volume tourism policy has worked extremely well for us, allowing the best economic benefit and the lowest impact on the environment. On a personal level, I have to admit to enjoying the sight of a huge male lion as he moves athletically across the grassland more than any in the bush. The symbolic beauty of a beast with so much latent power and enormous potential, the ultimate predator, in his natural environment, is hard to better. To see this animal living with such confidence of his ownership of an Africa we think of as our own is humbling. The humility of these moments is certainly not lost on Dereck and Beverly, as you can see in their words and photographs, a remarkable reflection of the place they love and have become so much a part of.

I personally hope that they never get comfortable, and that the hardship of getting stuck in the marshes and deserts of Botswana and living in a tent never loses its allure for them, because I look forward to many years of adventures together and discussions with them, and of course the next film or book.

Lieut. Gen. Ian Khama
Vice President of Botswana

Marsh lions used to the soft wet land underfoot follow the buffalo anywhere and everywhere around the island. No place is inaccessible, no place too wet or too difficult for them to hunt in.

Introduction

BESIDES THE MUSES AND SUNDRY GODS that conspire to put one in the right place at the right time, there were many flesh-and-blood people who pointed us toward Duba for some time before we finally got it. From the air, I'd fallen in love with Duba as I flew over it and its huge herds of buffalo. On the ground, we both saw its wonders and shivered at its dangers. It sang to our sense of exploration and adventure. But as an Okavango Wilderness Safaris concession it was forbidden territory, and we were too polite to ask if we could work there. One day we bumped into an old friend, Colin Bell, the Managing Director of Wilderness Safaris, who invited us to go down and take a look at Duba in the northern part of the Okavango Delta. We jumped into our vehicle the very next day and headed off, beginning a love affair with this isolated island that started with us getting terribly stuck for a day and a half in its muddy bosom.

We had underestimated the difficulty of working in this wetland. While I jacked up the vehicle, Beverly walked toward the nearest island to collect wood to put under the wheels. Suddenly she disappeared up to her knees and as suddenly came leaping out, scalded badly by what we found out was an underground fire, raging as they do for years and years in the layers of peat under the surface. We jacked up the vehicle and laid logs the whole of that day and most of the night, until 2 a.m. when finally we collapsed in a huddle in the back of our truck for a very uncomfortable four hours sleeping at a 30-degree angle. At dawn we rose to a most bizarre sight. Smoke and small flames danced from holes all around us as far as we could see, as some combination of pressure and temperature allowed the gasses and heat to escape in various places. Overnight the logs under the wheels had disappeared into the liquid earth below us. It was like being on the moon, and if anything should have chased us away it was that day.

This is a difficult and frightening place to work; to achieve anything one has to risk driving a vehicle through the most impossible conditions. Water flooded over the steering wheel each morning as we set out, and mud would suck the vehicle into the earth. I learned to drive through these crossings as slowly as the engine in the lowest gear could handle without stalling. Any added revs and the wheels can slip, and even an inch of slip can turn into a two-day experience. Sometimes our hearts were in our throats as we crossed 500 feet or more of flat water, waiting for the inevitable hole or wheel slip.

Our relationship with Duba and Wilderness Safaris started that day, and some ten years later Duba continues to enthrall us and fill us with the kind of passion we thrive on creatively. It is the place we call home, I suppose, if there is such a place for people who do the kind of work we do.

The search for this paradise is as much a quest for a real place as a journey to distant islands of the mind. This project has been both a physical and a metaphysical adventure for us, and indeed for anyone who ventures to a place were the very edge of life is revealed so unashamedly.

Some years later when we were ready to look at Duba again, we flew in with a friend who had come to our rescue nearly ten years back when we were so badly stuck on our first trip. James, a guide at Duba Camp, greeted us with such overwhelming hospitality that in answer to his question about when we were starting, we took a joint deep breath and said, "Today."

And we did. We were keen to get going and didn't know the tracks well, but we started out like six-year-olds with a candy budget. Around sunset the roughly drawn map was tossed into the back of the truck because we had no hope of finding our way back. Besides, we don't have "Let's go back" in our vocabulary. "Always forward." (Duba, however, has taught us these words!) Eventually, some time after dark we came to a water crossing that looked daunting (more so than the others). We had little choice, so I drove in. When the water reached my knees inside the car we were halfway. No going back. I kept the revs low and moved forward. As the water reached belt height I was getting cold, and getting cold feet. When it reached halfway up the steering wheel we were miraculously still plowing forward, but then we hit the deep water. The hippo path did us in. We were mostly underwater, and while I cursed and froze (it was the middle of winter and possibly four degrees Celsius), I kept the engine running and racked my brains for a way out. Beverly handed me the radio and plied me with Rescue Remedy. I knew the time had come to swallow that "get stuck" pride, and I made the call: "Eh, James!"

A Tsaro lioness ghosts through the dawn, heading toward the buffalo.

WHY DOES ONE want to watch lions and buffalo interacting in real life, view a film, or in this case, read about it and page through the images in this book?

Why watching the very edge of life and death, as one does here on the African plains, is important is a question that bears asking, at least. Why it is so thrilling is another.

Has our drive to witness Africa's harshness come about as a result of the titillations of too many NATIONAL GEOGRAPHIC magazine articles, or films on claw and tooth that make good television (or good television ratings at least)? Or does it stem from something else?

I believe that people who are interested in learning more about themselves and finding a solution to being must attempt, at least once in their lives, to be in the presence of the wild and in isolation from the clutter of modern society. For centuries we have sought to understand how to fit in, and why we should. It is still a mystery to most people in the world, and certainly to most influential people, but out here in the wild at least the journey can begin. Journeying to ancient places provides us with the clearest point from which we can at least start the internal process of discovery.

Although we may still find vestigial evidence of what we may be seeking at the heart of our various civilizations, confusion overwhelms us as layers of social myth and its intricacies litter our way. Only in a void can we exercise a clearer thought process. Quiet open space is not only our escape, but a necessary treatment.

Very few places will put you in the direct path of confronting your own life and death and mortality—in addition to the inner questions left too long unasked—as the wild places of Africa do.

Are the answers to these questions that we have been reaching for over the ages—questions about the fragility of life, the acceptability of death, and predation: killing and our unique place within that, or outside of it—to be found along this path?

There is little doubt that, as a species, we have somehow skipped out of the normal evolutionary processes, elevating ourselves to a position of control rather than being controlled in so much of our lives. Our weak are no longer eaten on the streets behind us, and we no longer (openly) turn our backs on those who fall behind. We live by a new set of rules now, rules we have made up ourselves. As a species we have excelled beyond the normal constraints of nature in our upward population growths, aided by medical care and a unique ability and desire to mate uncontrollably. In doing so we fill into other species' niches like mercury, filling into the smallest cracks and crevices in anything around us, mostly at their expense or demise. We can alter the very climate that shapes the environment that made us develop or evolve into who we are today. We can change anything and everything we want to, and some things we change whether we want to or not, just by being what we are.

Perhaps when we watch the lions stalk in and bring a powerful buffalo down to its knees we see a little of the complexity of the system into which we were born and from which we have extracted ourselves. And when we do, we must know that we still have enough of it within us to scare us. Despite the controls we add to ourselves, a remnant of that wildness stays within us, and subdued versions of it sneak out from time to time in anger and frustration. We can see beyond the action of the chase and the hunt, and understand, with some reverence for the moment, that this is as close as it comes to the edge of life, and for some even exhilaratingly close, not because of the blood spilled but because so seldom is one placed exactly at such a meeting of past instincts and present intellect.

If it weren't for suppression and self-control of course this would be a lawless life. But seeing life taken in the wild is a reminder of what it was, and what it could be. Could be? Well, because the same rules that we have imposed on ourselves that have so elevated us from evolution come with a responsibility.

Observing how it is for every other species that we allow to exist on Earth should stimulate us to even greater heights (beyond Mozart and the technological developments of our day) to consider what it would be like if we brought about ecological collapse. For while there is nature, we have the luxury to be humankind, and above everything. However, without a healthy environment, we may find ourselves under so much pressure to simply survive that, ironically, we would become animals once again. The rewards of superior intelligence have their dual burden of responsibility.

The balance, however, is the understanding that while we are leaning toward being apart from nature, we are in fact as linked to it as the lions are to the buffalo they hunt. We are as much a part of it as it is of us.

Perhaps as you page through these images they will give a sense of that part in you that refuses to allow the link with the forces of nature that made us who we are to be broken.

Remnants of last season's hunt sink under the rising winter flood.

FOLLOWING PAGES: *Lionesses scatter the herd into the deep water; one chases while the rest stay behind.*

The Place: a Map of the Heart

GEOGRAPHY TURNS SPACES INTO PLACES AND THAT MAKES US FEEL SAFE.
Yi Fu Tuan

DUBA PLAINS and the series of islands around Duba were formed, like many islands in the Okavango Delta, by the constant weave of the myriad channels and rivers; its annual whimsical flood; and the miraculous work of termites, who build their castles first as large as a vehicle, then a house, and finally, after years of breaking down and rebuilding and change in the acid content in the ground, an island. Floating papyrus, often broken away or discarded by hippos and elephants, ends up lodging against a small inlet and stopping. The water flows under it until it grows and often sends down roots. The channels change, and in time there is Duba, a group of Phoenix palm islands with a few dozen warthogs, 2,000 lechwe, 14 tsessebe, the same number of wildebeest, some kudu, baboons, a scattering of other new residents like aardwolf and hyenas, a couple of leopards, some elephant. And although that sounds like quite a lot, there are very few other local residents—no impala, giraffe, or zebra, for example. Oh, except the lions and the buffalo.

At Duba, the buffalo move in a set pattern that gets interrupted regularly and reestablished. The lions stick to their territories plus a small overlap, waiting for the herd to circle into their hunting domain. The buffalo use every trick they can to disappear and deflect attacks; each day is a race to minimize the damage, but each day there is damage. Well, most days really. On average we record 15 kills a month. Most successful kills are preceded by multiple attacks and attempts that take the lions on hunts through the water and through the midday in long, relentless chases that eventually wear the buffalo down or wear the lions down. It is never certain who will be worn out first by the process, but each day this goes on and on and on. It is why we call this project Relentless Enemies.

The scene has been set. Those are the participants. I won't call them players, as we would use the word to refer to the actors in a theatrical scenario. The word "players" implies play and fun, which is completely the wrong sense of this drama. This is about death, whether our own or an animal's, and these are important moments for everything around, including the observer. Understanding more about the hunt and the kill, as well as our own feelings about life and death, is what this is about. Duba has become a special place, a spiritual place for us and for many who visit it.

Africa is also the place where man was born. When we visit again the house in the suburbs where we were raised, or the apartment we first grew up in, a certain attachment washes over us as the distant memories, now foggy with age, struggle to the surface.

So it is with Africa. Those distant memories claw their way out of our deeply hidden and disguised primitive past. We experience these "feelings" before we can intellectualize them, and so label them as our primitive soul because they come from inside. But these are ancient and perhaps not too distant memories, and if we accept them and let them take us on a journey, we will find great riches of understanding in our African origins.

Masters of the underwater in the same way as lions are of the open grasslands, crocodiles are also a symbol of Africa.
FOLLOWING PAGES: *Space rather than place gives these plains their life, and their allure.*

Clear tannin-leached water does little to hide the hippos and crocodiles in the river to the north of the buffalo range.
OPPOSITE: *The Ngoga River slithers through the watery papyrus desert that forms the southern boundary of Duba's buffalo movement.*

All inhabitants here must learn to contend with the deep rivers that surround them, and all they represent.

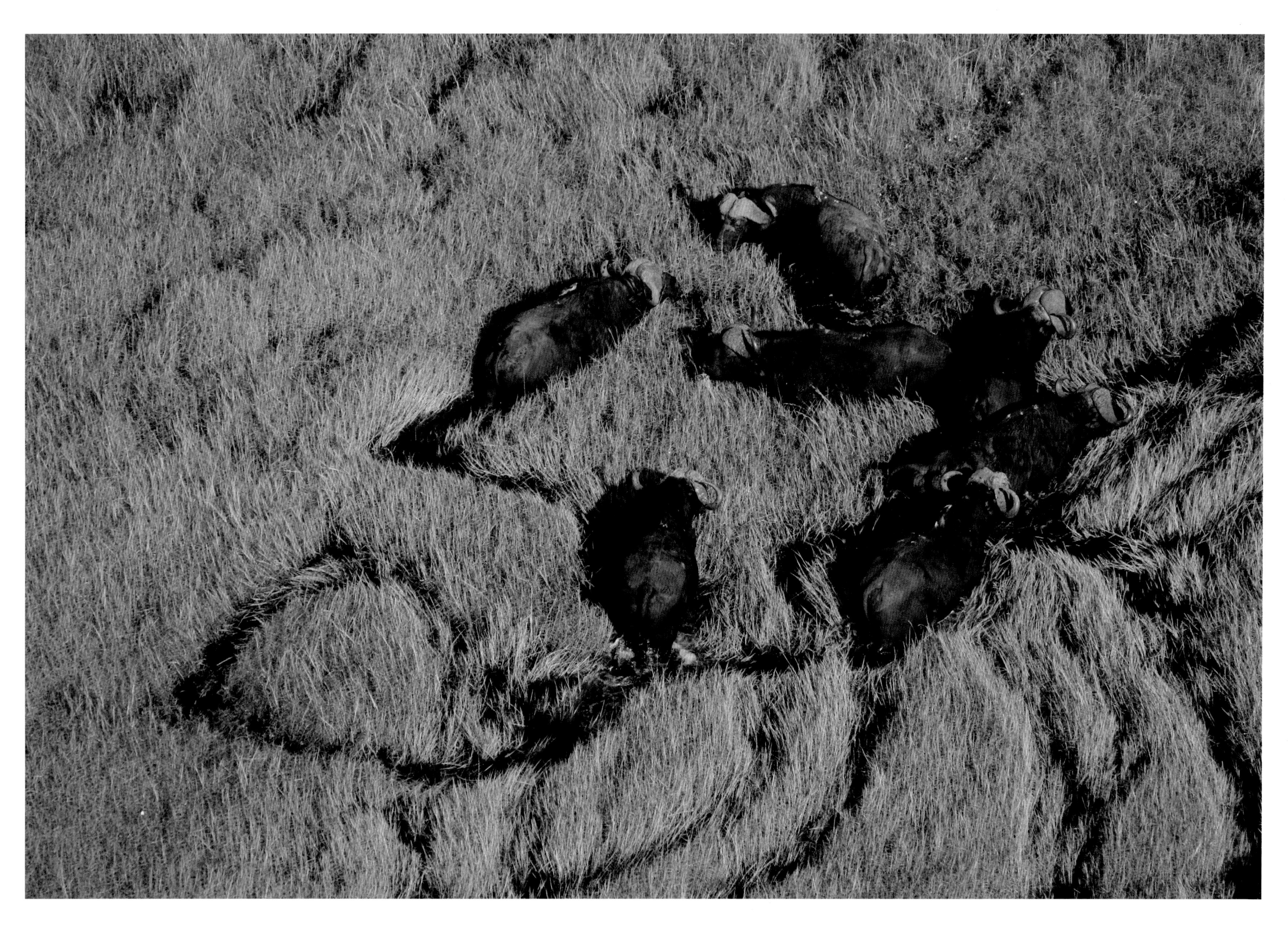

The old bull buffalo vacuum up the soft grasses and delicately avoid head-to-head confrontation at the same time.

The herd, over a thousand strong, marches from one end of the island to the other, avoiding as best they can the ravages of the lions, but staying within the boundaries of the ring of deep water that surrounds Duba.
Opposite: *Hunting in water for hours each day has forged heavily muscled chests, necks, and upper legs, making these lions the biggest we have seen.*

At the water crossings the lions force the buffalo herd to move faster, using the drag of water and the panic of confusion to select a victim.

WITHIN THIS ANCIENT RELATIONSHIP, a new, organized, and systematic trend among the lions has emerged. Lions are the great communicators of the bush, and communication is what holds all social structures together. Somehow, perhaps in the same way that two people find a common purpose just by being together for a long time, the lions of a pride seem to reach a level of knowingness without seeming to communicate visually or vocally. They move into a hunt as if they have a map in their hearts that each knows, a map of each hunt, a map of every move that they have traced either in their minds or by experience time and time again. Lionesses leave or approach the buffalo as one, silently, and circle around the herd to a position they seem to know, and into a pattern of a hunt they obviously understand. The strategy has worked before, and the more they slip into the pattern the more they succeed and the deeper the strategy is etched into their consciousness. And all of this happens without a sound from these ultimate communal hunters and social predators of the wild.

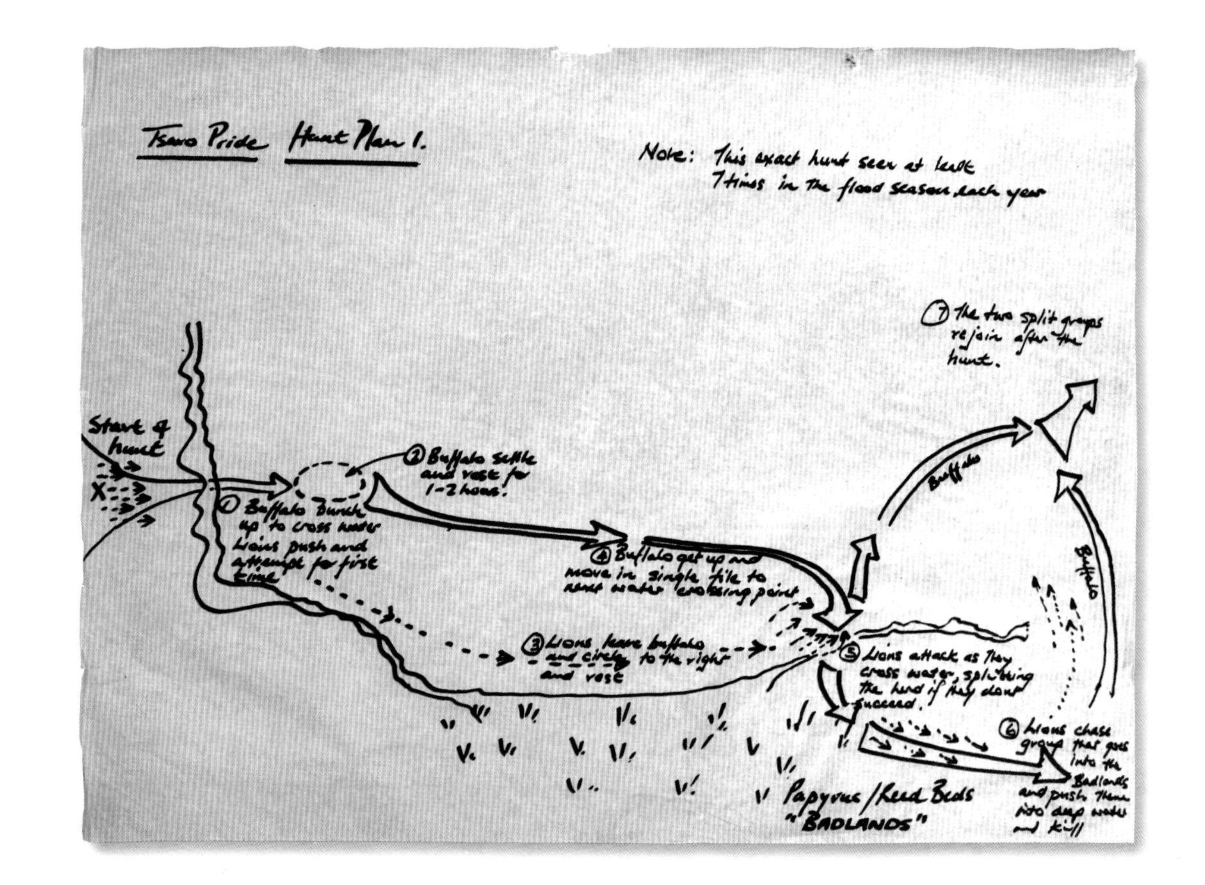

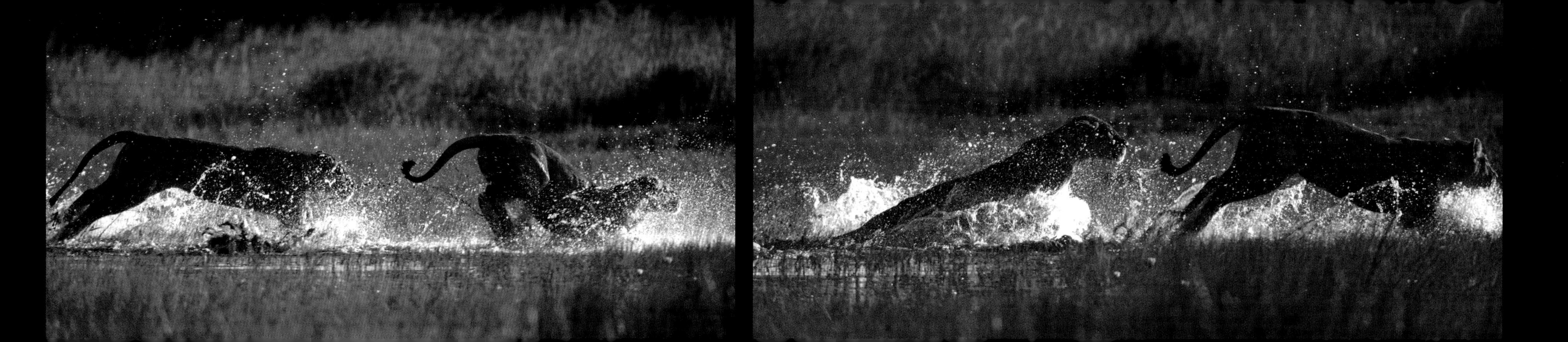

A swamp cat at work, endowed with heavy triceps, neck, and chest, is propelled by thick sprinter's thighs and lithe lower legs.

The Eternal Dance

Quiet dream of brief repose above the world, the sun blazed, the rock glowed, the tree rose unyieldingly, the bird sang harshly: Eternity! Eternity!

Hermann Hesse

TIME AND SPACE HAVE FORCED these two animals together. Their dance is eternal, their destinies intertwined like one beast, a double-headed shape of amber and black, working together like muscles in the same body.

They dance again and again, each day, each to the same rhythm—shapes in the grass, closer and closer, an explosive snort of alarm, drumming legs, heads up and charging. The shapes duck away and disappear, the grass waves gently in the wind. Too many smells float around, confusing the direction of death. But even with the best plans and avoidance, the shapes hunt on.

In the end, someone dies. It is the way of this place, the way of Africa, where there are simply those who hunt and those who feed the hunters. It is lions and it is buffalo and it is Duba.

Wars are not fought for survival. They are fought for ethics or religion, ideologies, space, or resources. So to associate this relentless interaction between the lions and the buffalo in terms we can understand as war is erroneous. It is definitely one between predator and prey, and yet there is a beauty in its perfection. There is a playfulness in its execution in a way that parallels the heroic battles of mortals in the Greek tragedies, as entertainment for the gods. Unlike the Homeric legendary battles of Troy and the Iliad, this dance illustrates an interdependence as much as it is a conflict. Although the benefits seem one-sided, the buffalo thrive under the pressure, in a Darwinian sense.

This eternal dance is by definition "forever," although that assumes there will be dancers in eternity. And as we stand back and watch this poetic shuffle, I cannot help but feel somehow sad at being left out. Throughout our own history we have danced our own dance with the great predatory cats, over millions of years, and it has taught us most about ourselves, more than many of our other experiences have. And while the biggest lesson we could learn was to survive, which we do amazingly well, it has extracted us efficiently from the dance itself. At our best we place these big cats in cages or sanctuaries. At our worst we wipe them out. David Quammen (author of *Monster of God*) suggests that within 150 years there will be no dangerous wild animals, like lions and crocodiles, simply because we won't allow them to exist. It certainly seems like that to me as I sit through endless debates about how best to accommodate our lust to kill them, or which populations of lions we can still save, or where our efforts are best spent. It sometimes feels like planning what furniture to save on the *Titanic,* instead of looking out for icebergs. And sometimes I close my eyes and remember the dance that still rages in my soul from four million years ago, and miss it.

So the eternal dance is theirs and ours.

The back-and-forth interplay of power between two of Africa's giants is eternal, harsh, and at the same time quite beautiful and essential.
FOLLOWING PAGES: *The dance is relentless: The buffalo attack the lions, who dive for safety only to bounce back into the dance immediately.*

IT SEEMS AS THOUGH it is not only individual buffalo darting out of the way that feels so much like a dance. The entire herd sweeping away in a set pattern ahead of the whole lion pride resembles a dance on a grander scale. These patterns so parallel the well-learned and rehearsed steps that make up a fluid dance that it certainly feels as though these lions and this herd of buffalo have been doing this forever.

The map or diagram below is another example of a hunt plan that the lions seemed to just slip into once the buffalo reached a certain point: breaking away from the more obvious tactic of following the herd, to sweep around into ambush, strategically, clinically, and using forethought. It has been thought that the qualities that distinguish us from animals, or other animals, are our amazing ability to be self-aware and our grasp of past, present, and future. All animals, it was once thought, have a grasp of present, but past (memories) and future are beyond them. But these hunts, like so many behaviors, indicate that animals have a very well-developed sense of the past. In this case, they know that moving ahead to the water crossings brought them past successes. At the same time they clearly know about the future; again in this case the lions know that by going ahead they will (in that future) get into position to split the herd as it crosses the water, and increase their chances of success. Each time we set a new level to distinguish ourselves from other animals, we find out something new about them, new information that makes us raise or change that bar. As we do, the lines between "us" and "them" blur. This process leads us quite quickly to the unsettling arena of thought about animal emotions and, even worse to many people, animal souls. There is little doubt in my mind that animals have emotions. It would be most strange for one ape to have developed emotions when others, in fact all other animals, have not. Emotions have armed us well for our development and survival. Clearly all animals have feelings like fear, a very useful emotion, so why not joy and even love?

Perhaps the mistake we make is not in playing down their emotions by saying, "Well, they do it for some evolutionary function," i.e.,"play helps create a bonded pride that hunts better together and that equates to food all around," but rather in elevating our own emotions beyond the rules of nature and function. Perhaps we, too, love and experience joy for very simple and basic evolutionary reasons that serve us in this short stay here so that we can survive to better continue the species.

These are cerebral as well as physical hunts, a blend of thought and action, a mind game resulting in a kill.

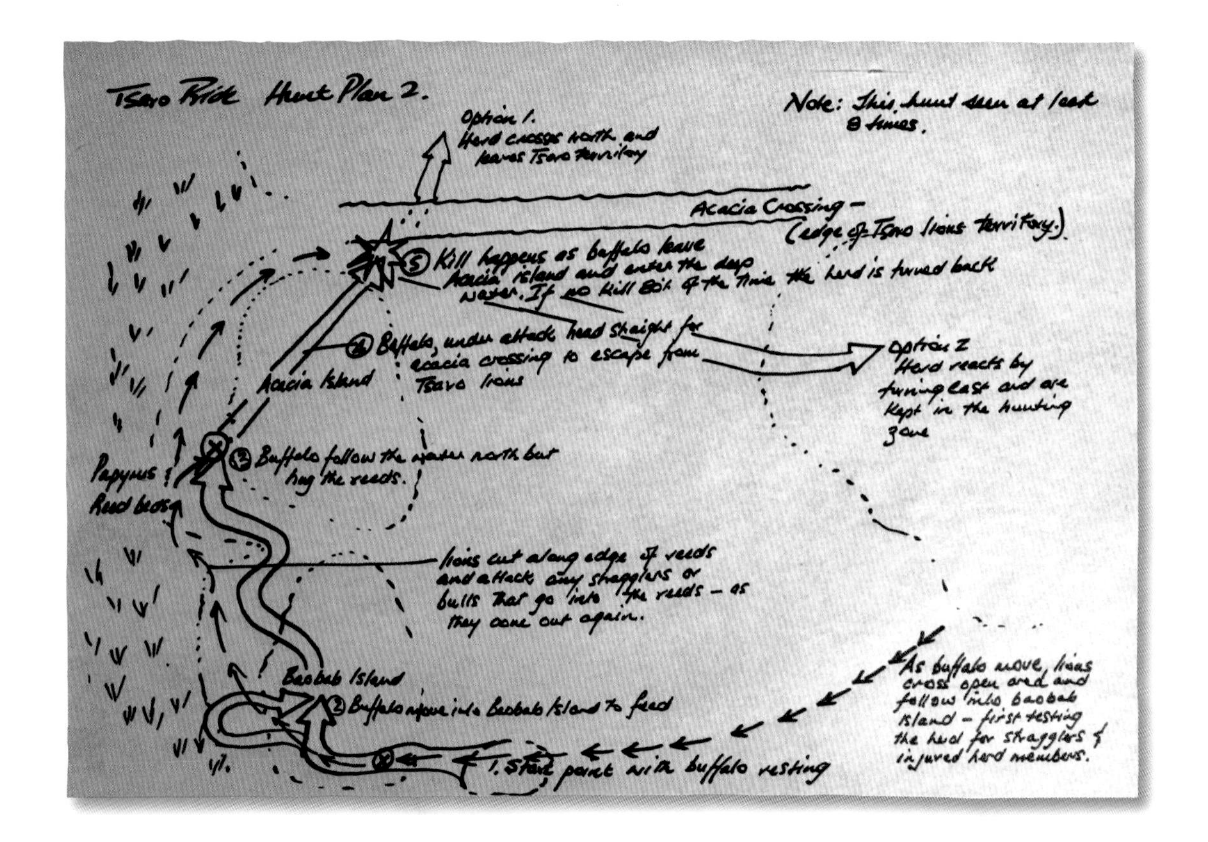

Darting for safety, running for their lives, dashing for cover: The lions in Duba are always at risk around the buffalo.

Both bull buffalo and cows show their disdain and displeasure at being followed. Elaborate defenses, irritation, or avoidance tactics go on for hours some days.

The buffalo break out of the herd and defend aggressively, then quickly dart back before they can be isolated.

Lions: Shadows in the Dust

HAS THE LION COURAGE? WHAT IS COURAGE?
Sir Alfred Pease, 1913

FROM OUR EARLIEST MEMORIES of lions we have held them up as icons of what we see ourselves as being (at our bést moments): noble, brave, muscled, and social. Their biceps and triceps resemble ours, and their group size and structure is similar to what paleoanthropologists believe was our ideal group size in our first forays into the savannas as early man.

We have erected statues to their image and glory, and we call our great kings and leaders by their name. In Botswana the president is referred to as Tau e tona (the great lion). Richard the Lion Hearted speaks for itself.

Similarly, another modern icon is the large bull buffalo, often considered the ultimate herbivore. Like many of our symbolic idols, we value this animal because of its ability to kill people with a fairly large statistic. We seem to idolize the very things that we spend millions of years avoiding being killed by. It is the "noble enemy" relationship, which emerges from within us all from time to time. In addition, this strange evaluation of buffalo is a hangover from an old hunting era, because it is said that they are difficult to kill. It can be tough to hunt them, I imagine, if you are armed with a bow and arrow or stone tool. Today's rifles demystify that; shooting a buffalo is quite easy now. But when a bull buffalo in its prime snorts at you and comes crashing through the palms at close quarters, slashing and smashing fronds as it comes, it most certainly gets your attention.

If you ever get a chance while out on safari one day, try to lift a set of buffalo horns over your head. Your arms will strain with the effort. These buffalo wear these horns and wield them with the dexterity of a fencer using his foil, but on necks built not only to hold up those horns all the time, but also to take the battering impact of an opposing male who will drop his head and charge with the speed and intensity to knock his opponent senseless and off his feet.

Sometimes in these battles with lions a bull will drop his horns and present a gnarled beaten boss, that solid dark battering ram that protects the skull, and charge. The lionesses know that it would be a bad mistake to take this on directly. Images of a puffing steam train often rise to my imagination as the buffalo kick up dust in a misty sunrise and charge at the lions.

These two opponents are locked in a relentless interaction, here in this place and in this time. But so much can change. What happens in Duba Plains, this unique set of islands in the Okavango, will be different from what happens in East Africa, or southeast of here in South Africa. We have known some areas where the lions balked at hunting anything as dangerous as a buffalo. At Savute, in the dry part of northern Botswana, where we spent 14 years recording lions' hunting behavior, the lions so preferred hunting buffalo that they would literally stop any hunt halfway through at the slightest sound of a distant herd of buffalo. However, they hunted almost exclusively (more than 90 percent) at night and ceased all hunting as soon as the bright moon crept up into the sky.

This is not the case on the plains of Duba. All animals may have a baseline of characteristic behavioral patterns, but they also have a large layer of cultural, time- and place-related characteristics. Some lions just learn how to hunt elephants and then specialize in doing so. The Duba lions, as bold as they are with buffalo hunting, have never killed an elephant.

Lions seem to test and learn in much the same way we do. We are best at finding patterns, fitting those patterns to our own past experiences, and learning from them. In Savute in the 1990s, when the lions started hunting elephants, it seemed as though they tried once or twice; then when they succeeded, it was as if a floodgate of knowledge was opened, and they hunted larger and larger elephants.

At Duba, the lions have learned what works for them. In time something will change, and they will learn something else that works better for them. To lock them in permanently as being daylight buffalo hunters, as they are now, would be a mistake.

Masters of all at Duba, the males are probably brothers, aging now at 13 years.

The males in Tsaro pride are greeted like adored pride members, unlike many other males, because of their long tenure. When there is mating, which is still very common and successful despite their age, the males are aggressive.

The males in this pride seldom hunt or even participate, but they always dominate the first feed.

Tsaro Pride: Confronting Death

You whose roar frightens the dreams out of the heads of sleeping warriors!
Ngonyama, you, whose footprint is never erased by the passing winds!
Credo Mutwa

Nine lionesses and an ever varying assortment of cubs; these lions always seem to have small cubs, or they are pregnant or mating. The pride has an unusually high cub mortality, which at first I thought could be because the males are so old now (at least 13 years), creating a sense of pending instability. Or perhaps it's due to disease, bad motherhood, the swamp itself, or the buffalo in some way. It puzzled me for a long time. For three years in a row cubs have been born copiously: 22 one year, then 15 the next, none of which made it to six months. The number of cubs rules out any sexual disability of the males. There may be some genetic deficiency involved, and certainly the buffalo may trample one or two cubs a year.

The swamp itself does have its own character, and it can be a dark one. Beyond sucking our vehicle down all too frequently, it houses crocodiles, and in some cases barbel (catfish) large enough to snatch up a small lion cub. Bad motherhood is certainly a possible cause of the small survival rate of the Tsaro pride's cubs. We have seen females walking miles after the buffalo, letting a line of cubs fall farther and farther back, but marching on without too much concern. However, in May 2005 we saw an incident that alerted us to be more watchful, and as we watched a trend developed.

Out of the corner of my eye I saw her lean back tensely. Then she suddenly stood and pounced into the grass, raked her claws in quick succession through the grass, then dipped her head down, teeth bared. All of this happened in less than a second or two, but in that time she flicked one of the four-month-old cubs at least three meters into the air. It was dead when it landed, guts spilled. Lionesses ran everywhere, some away from the scene of the crime, some toward it. The other cubs ran off across the plain and kept going. We noted her ear notches. She was a cubless lioness.

The pride was in disarray for hours, lionesses calling softly and approaching one another cautiously. Some walked over to the grass and sniffed. One called at the dead cub, and they all stayed around the spot the whole day, even though the buffalo moved off. Then in the afternoon one lioness approached and sniffed the cub, called and sat down, then slowly stripped off the skin and ate the cub. Stunned, we took our pictures silently and then backed away as if we'd seen something private and illegal. When she stood and walked away, her suckled nipples showed her to be the mother. Half a dozen times after that we saw the lioness that had killed the cub lean forward, ears alert, looking at another cub. Twice when her neck arched, ready to pounce, a mother noticed the posture as well, growled a warning, and collected her cub. Within the privacy of their nights that week, five more cubs disappeared. The pride wandered around after the buffalo but didn't make any kills for the whole week. They were a disturbed family. Then the small young female with clean ears (no cuts or nicks in the profile of her ears) with two new cubs decided that it was time to introduce them to the pride. We could see them coming across the grassland, and so did the cub killer!

She looked up at the three distant figures struggling through the grass and broke away from the buffalo standoff that was going on up on the high ground. She walked the 15 minutes it took to get there like a lioness going to greet her companion, just a normal social bonding. But then she arced around the female and started trotting toward the cubs, ears forward, neck arched. The cubs dropped down and cried. This was their first encounter with any lion besides their mother, and it was bearing down on them fast. The more slightly built mother had been outmaneuvered, and the cub-killer (possibly her own sister) was between her and her cubs. But in an instant she was right there, on her sister's back in the dust, in an attack delivered with such ferocity that the cub-killer collapsed in submission and balled up against the barrage of blows.

Thus stabilizing the situation, the mother called to the departing cubs and stopped their retreat. As her back was turned, the cub-killer relaxed and lifted her head. She was rewarded with another set of ten or so swipes and bites delivered so fast that it was over before the dust could rise. The female turned on her back and submitted, effectively stopping the attack, and at the earliest opportunity crawled away on her belly to a safe distance where she pretended to go to sleep for the next half hour. The cubs watched with big eyes, secure for now. But the Tsaro females have lost 39 cubs in two years like this. Possibly it is a natural reaction to the confines of a small island, and perhaps they have no other way of controlling numbers, but it is a harsh method of keeping stability.

The Tsaro pride, two huge lions and nine huntresses, dominate the heart of Duba.

Silver Eye, as we called her in preference to the Blind-eyed Female, took on a certain beauty once we'd named her. Lions are supposedly all of equal rank, but Silver Eye stood out for more than just her features. She is an excellent hunter, always taking risks either as the first female to attack or among the front line. Perhaps that is how she lost her eyesight; a kick or attack by buffalo could have been the cause.
OPPOSITE: *The other large-bodied Tsaro females carry the typical remnant stripes on their backs. We'd seen this in Savute before, but it isn't evident in all lions. I suspect it's a ghost stripe from some past time when lions had stripes.*

Tsaro cub production hasn't been good recently. The long water crossings are difficult and dangerous. Crocodiles, drowning, and simply falling behind may be the causes in some cases.

MY BEST GUESS is that all of the difficulties may add up to whittle cub numbers down naturally. This island wouldn't have much left on it if every cub born survived. Ultimately, just as lions have a lifespan, so do prides of lions. We have seen whole prides simply die out. This happens when more lionesses die than cubs are incorporated into the prides. What we have at Duba with the Tsaro females now is a pride that may have a traditional (or ecological) ideal size of nine females. There is no room for more lions in the pride right now; the excess will die or move on. Right now they are dying young.

Nevertheless, the Tsaro pride is flourishing. These are the largest lionesses I have ever seen. Thick necks and bulky bodies distinguish them from the other prides around. The reason may be their almost exclusively buffalo diet, but it may also be the exercise: constant following and then big attacks on large prey, which often need wrestling to the ground. Either way, these lions are huge.

Their special hunting technique is different from that of the other prides, and I call these lions "The Confronters." They seldom stalk in, but walk in boldly. They usually start the day by walking up to the herd to get them on the run. Then they scan for any weaknesses.

As the buffalo herd runs, the lions move their heads as if paging through the telephone book, looking quickly for the telltale sign of a limping animal, young or old—preferably young. The buffalo run, but then return to chase the lions. The lions wait their ground, perhaps as a test to see just how far the buffalo will push and be pushed, each move being watched by the lionesses. Then at some point they may chase in directly to get the herd running, then follow through with an attack.

This is confrontational and relentless hunting. Some days the hunt has gone on for six hours. The lions' technique is to walk in exposed, push the herd, avoid the counterattack, follow back in and push the herd, avoid the counter, find a weakness—patterns, patterns, over and over again, until someone makes a mistake, a calf is left near the back, a limping cow struggles to make a water crossing.

Recently the lions have had success in attacking from the side, splitting the herd and waiting for the stragglers or a smaller herd to be nervous enough to want to join up with the main herd.

A classic hunt, one that we have seen over and over again, is a good example. The lionesses pushed the herd and then caught them in a long running line, from the side. Then they attacked and split the herd. After that it was a waiting game.

It was an hour before the two herds started working their way toward each other, but one female buffalo suddenly realized that her calf was with the other herd and broke away, swimming the river past the hippo pool and out onto our bank—the bank with nine waiting lions.

Tsaro pride females seldom look at the herd as it moves forward after the initial morning testing, and it is strange to watch them all turned away from a thousand buffalo. But they are looking for stragglers, and even the splinter groups of bulls that lag or weave off to the flank of the herd draw their attention. They seem to know that at some point these splinters will try to rejoin, and when they do, if the lionesses can get between them and the herd, the hunt is on. Buffalo always seem weaker as a splinter group and try to join the herd, often in panic. When the herd is nearby you can predict the route this offshoot herd will take to get back to the main herd. So can the lions.

At a certain time of the year these lionesses hunt calves almost exclusively. As the birthing season starts the lionesses change tactics. They push the herd until the calves drop back, then they run in and collapse the calf just long enough to do some damage before the herd returns to rescue it. If it is a long hunt, the herd may leave, and only one kinship group stays behind to defend the calf. At that point the lions want to get rid of the rest of the buffalo so they can eat their catch in peace.

Many hunts, however, have ended in two or more kills: lionesses see that a mother may be vulnerable as well, and give chase again even though they have one buffalo down already. And for some reason female buffalo collapse almost instantly. An animal that size should be able to put up a stronger fight. Even impala battle their attackers harder, in my experience. Male buffalo stand and fight for up to an hour, yet a female that is only 20 percent smaller than a male falls in seconds.

As a result of losses, the Tsaro females breed out of synchrony at the moment, so larger cubs and small ones mix in together, making it tough on the smaller ones and making new mothers more defensive, often with fatal consequences for the older bullying cubs.

Cubs in the Okavango have to very quickly adapt to water and conquer their innate dislike of getting wet. It is a difficult lesson for a cat, but from the first days the Duba lions take to water easily.

One day the tension between mother and cubless sister exploded into a full lion fight when the sister ran toward some young cubs about to be introduced to the pride for the first time.

The cubs took fright but the mother intercepted and attacked the female, whose intentions were clearly to harm the small cubs.

To watch in the early morning light, as the nine lionesses, almost identical females, fan out and start to move in on the herd, is enough to take one's breath away. The sheer beauty of this moment always gives me cause to breathe in and fill my lungs with life. It feels like the moment before some great event, a storm perhaps, a reading by the Dalai Lama, or a birth. Its purity is of a type that is so clean that you want to capture everything about it. You want to freeze the moment in a photograph, symbolize it in a painting, detail it in slow-motion film footage and sound: a crystal so perfect it needs its own showcase. It is that moment often reached for in stories and films, when out of chaos something suddenly synchronizes and is whole, the moment when a symphony, Mahler perhaps, rambles and bumps and then quietly finds its harmony and becomes one, not an ensemble of players, not nine lionesses, but one melody. And you know it isn't folly. Someone is about to die.

I don't get excited, nor do I get as anxious as others do when they watch the start of a hunt. But something fills my body with the profound—and everything seems to sharpen. Perhaps it's my need to be professional and take observations, as well as get into the right position without interfering, and then film it all. So we are busy, but the sound of grass brushing against their faces flicks in my ears, the crunch of their feet is agonizingly loud. An alarm call by a francolin indicates that the male is coming in (and potentially blowing the hunt). The wind drops and flutters a piece of hair into my face; it has changed direction and the buffalo will pick up their noses soon. They do, and they flick their tongues into their noses to add moisture and enhance their sense of smell. They know, as we do, that the time has come.

After that it is all mechanics, the job of getting it right, for both the lions and us, the magical moment over but still burning inside.

The Tsaro pride have learned a few other things.

The buffalo use the whole of this island and cross over to it in three places. But the Tsaro pride live only on this island, so the trick is to keep the buffalo in their territory. Some would say that forward thought like this is beyond animals, and is a purely human trait. Tsaro think ahead and turn the herd back as they head for the river if they are going north. Once in a while the lions' hunger drives them to attack as the buffalo are crossing, and they lose the herd for days at a time, as the buffalo swim toward an area called Paradise and an adjoining lion territory. But most of the time they ambush the herd just before it crosses the river, effectively turning it back home.

On the other sides of the island the lions know that they can push the herd as much as they like, because in the south an impenetrable bank of reeds and papyrus will eventually stop the herd and send it bouncing back—into the waiting lions.

To the west, in an area we call the Badlands (because it is bad: broken springs and bent chassis are the reward for attempting a crossing for us), the buffalo struggle over uneven ground and risk a strained leg or other injury. The lions will go in after them, pushing the herd to water and through difficult terrain, and then sweep for the injured.

To the east, the herd can be followed, but there the lionesses always spend their time aggressively looking around for the rival Pantry pride. (Not names given by us, by the way. These lion prides have been named by the guides at Duba Plains camp, the only safari operators in this area.)

Soft mouths for carrying cubs are also hard mouths for killing buffalo.

THE MALES OF THE TSARO PRIDE, known locally as the Duba males, bear mentioning. When we first got stuck in the mud here it was New Year's Day 1996. At the time Wilderness Safaris had recently taken over the management of Duba Camp (which they still manage), and records of the lions there have been kept since that time. The Duba males have been the residents since then, and have been the dominant males of the Tsaro pride. To become dominant, they would have needed to be at least five years old. That makes them at least 13 now, almost a world record as far as I know. We once followed a male who was 14 when he keeled over and died in front of us, but there was no mistaking that he was on his last legs at the time and had not had a pride for at least five years before that. He certainly wasn't producing 22 cubs a season!

These old Duba males are still mating and fighting over food. One has a broken tooth, but other than generally looking a bit tired and battle-scarred they are like most other prime territorial males. Much has been written about a lion's mane. It is a burden in a hot climate. Some suggest that a black mane makes the male at least 12 degrees hotter. The mane is unlikely to have evolved as a protection against attack while fighting, since females fight almost as often as males do, even over territory. But it is a very visual signal that often works against the whole pride when the male wanders along to meet up with his hunting females and scares off the prey.

As a signal to distant potential challengers, however, the mane is a very good banner. It won't prevent a fight; if the challenger wants to fight for territory he will, but the mane certainly signals that a dominant male is in residence and prevents a challenger from thinking the coast is clear. The Duba males have reasonable manes, one dark and one lighter. Their faces are like old carpets, bitten and beaten from too many years of fighting over food and females. But old males who are ousted often lose their manes almost immediately; they dwindle down to scraggly limp dreadlocks that won't attract the attention of any aggressors.

With the average tenure of males being two years, these males' ten seems unusual. Why? Well, one answer is that there are no challengers, which I can confirm is the case here. The last potential challengers wandered into the area four years ago, and were quickly seen off by the Duba males in their prime. When a lion roars on this island we know who it is because we have either just left them or we are with them. We know all the lions here. When I see a track I know which lion it is, and usually where it is going or coming from. There are no strangers here. We would know. Tsaro pride would know and would react. And to discover why we had to look back in time.

Prior to 1995 this area and the adjoining ones were trophy-hunting areas. We had seen in Selinda and other areas we lived in that the hunting of lions can easily spiral out of control. The shooting of one male can result in up to 30 lions dying because of the structure of the pride, the coalitions that males form (if one is killed his partner is left vulnerable), and infanticide by new males. Trophy hunters seriously damaged the lion populations in the northeast where we lived, and we watched the collapse.

I don't know what level of hunting happened at Duba; however, with each hunter needing at least one male lion from this area every 21 days for the hunting season, I can imagine. Every male was shot, prides disrupted, challengers of young age shot as well (as they were in Selinda and Linyanti), until the pool was dry. How these males survived that era is a testament to their wiliness, and also an indication of their age. They must have been just too small to be shot in 1995, perhaps three years old. However, something happened one day while we were sitting with the pride earlier this year to indicate that they were at least aware of what was going on around them as young cubs.

At the camp, at least three five kilometers away, the manager accidentally set off a "bear banger," a loud gunshot-like explosion. The lionesses (who must be much younger than the males) didn't flinch, but the two males leaped up and stared in the direction of the shot. They slunk off into the thickest thorn bushes Duba provides, looking alternately at us and back in the direction of the camp. They remembered!

I know that many people feel the need to shoot a lion to prove something, and even more people don't have feelings one way or another about that need in others to do so. I have some difficulty being ambivalent. I have an even harder time liking those who need to shoot lions. I don't like people who need to show off their prowess or their egos, and trophy hunting stems from the same need to boast and show off, in my opinion. I think that our society deserves better of our alpha males. We need men, and women, who can now lead us into a new era that offends nature less, takes from it less often, and thinks less selfishly. Thankfully, that time is coming.

Against a constant backdrop of buffalo, the lions of Duba become highlights on a dark canvas, a part of the whole where two interdependent species are locked on an island together.

Skimmer Pride: Followers in Paradise

SECRETIVE SILENT AND SUPPLE AS A PIECE OF SILK...
M. Edney

BY MID-2005 THEY NUMBERED 14, including nearly grown cubs from their one (also old) male and four females. By contrast to the Tsaro females, which are so extraordinarily large, these females are small or at least normal in size. Genetically, adjoining prides are often related in isolated areas like this, but these lions just have not developed the bulk of the Tsaro females.

It was a drizzly day in summer when we found them hunting. The buffalo had escaped from the Tsaro territory and had crossed through the water, leaving the nine lionesses chest-deep in water watching their prey disappear. Earlier they had killed a calf in the water in spectacular fashion, spraying water everywhere and hauling the baby away from its mother's side. But one of the Duba males had commandeered the food, and they were still unsated.

The herd moved north. Having been stuck up to our axles the day before in water we call the Smelly Crossing, for obvious reasons, we ventured around through the edge of the hippo pools and got ahead of the herd—and found the Skimmer pride.

A lone female buffalo kept turning back and calling. She must have been the mother of the calf that the two Duba males growled and fought over. The whole Skimmer pride woke up and started to run in.

They stopped and watched the herd go by, waiting for the stragglers, and then finally went into their usual strategy: they followed. I call them "The Followers." Their special technique is to follow each move the buffalo make, just a few hundred meters behind, and to stay out of their way. Sometimes the buffalo turn back and confront them, and the whole Skimmer pride then heads for the trees, scrambling up to safety. It is unusual to see the horizon dotted with lions swaying uncomfortably in trees.

On this day, they eventually spotted a limping adult female. The four lionesses went into position and ambushed her. After days and months of driving through meter-deep water, our filming truck had long since given up any semblance of a braking system. As we drove up to the two females wrestling the cow to the ground I couldn't get the car to stop, so both of us grabbed cameras and let the vehicle shudder to its own halt some way off. We filmed and photographed the kill from the ground, which elicited only the slightest glance from one of the females, now intent on getting her prize off its feet and onto its back. As it went over we backed off, and the incident went off without a problem as the rest of the lionesses and the cubs raced in and killed the female buffalo. It was a typical Skimmer hunt.

The interactions between Skimmer pride and the buffalo take on a certain playful feel mainly because of the cubs, with buffalo snorting and chasing and cubs ducking and diving through the grass and doubling back to pick up the game as soon as the buffalo run by. The lionesses watch, doing their dual job of scanning the herd, getting to know it intimately by observation, and keeping an eye open for cubs getting into trouble. These are good mothers, and they have reared cubs successfully in the past few years. I suspect that the pride will take on the female cubs and grow to eight, or nine. The male cubs will be cast out to wander until they can find a territory of their own. At the moment there are five of these young males, and if they all make it they will be a formidable coalition. Maybe they will return one day to finally oust the Duba males (who by then will need walkers to keep up with their females).

The most amazing moments to watch with the lions at Duba are when they decide to swim, which they do with a comfort and willingness to cross water that is unexpected of lions.

Skimmer pride were the first lions we filmed crossing. "Comfort" may be the wrong word to describe their swimming, because they clearly don't like it, despite the fact that these lions cross water daily. In Savute, where we did most of our previous lion work, lions avoid water at all cost. One lioness there circled an entire dead elephant in the water a number of times before reluctantly dipping her feet into the water.

Skimmer pride, in contrast to Tsaro, are successful cub raisers for now, rearing nine of their ten cubs.

IN DUBA THEY CROSS A FEW TIMES A DAY. Some of the crossings are so deep that the females swim, paddling across and snarling an awful face to keep their whiskers dry.

Cubs do exactly the same, and we followed some cubs on their first water crossing. They smelled the water. One cub jumped back at the feel of it on his nose, then padded in before their mother. When it got deep, however, he had to paddle across, and he launched into the water with such certainty that his mother became alarmed and went in to get him. The others cubs, however, hung back until called, and when they went in they raised their whiskers like old women raising their skirts at the beach as seawater splashes over their ankles. When elephants swim, they hold their tails up and out of the water for as long as possible. For lions, it is their whiskers.

One evening we watched as the Skimmer females, as usual, followed behind the buffalo across the water. The evening light was just beginning to fade and we had half an hour left before dark. Suddenly one lioness looked over her shoulder and saw that behind her a group of about 70 buffalo had been left behind and were anxious to join up. She turned and stalked back, wading into the water toward the buffalo, who in their own way were wading into the water toward her. She kept going, a single amber figure walking boldly into an approaching wall of black shapes. Then, when she was about 20 paces from them, the herd started to run—toward her. I thought I saw a moment of hesitation as her neck stiffened, and I was sure that retreat was definitely on her mind. But she quickly conquered that thought, and launched into a head-on running attack. The herd turned, and in the chaos of mud and water this one lioness leaped on and attacked an adult female buffalo, bringing her down in a bundle in the water. It was one of the most confrontational kills we'd seen. Emboldened by this, the rest of the pride ran almost straight past the female—still struggling to kill the cow—and in the distance killed another buffalo in the last light of the day.

It is quite rare for these lions to kill at night or even in the last light of the day, and there has been much speculation on exactly why. It was a puzzle that I was determined to solve, and I asked scientists and lion experts as well as bush people whenever I could. In fact, I asked everyone. The puzzle is this: Lions have relatively small hearts for their body size, making them less efficient in the heat. Duba, like most places in Botswana, can get very hot, certainly not less so than Selinda and Savute, where lions are mostly active at night. Why should these lions hunt during the day?

An associate from Oxford University put it quite simply by saying that for a behavior to occur it should be a) a distinct advantage; b) the opposite of a distinct disadvantage; or c) possibly random.

Let's say an advantage might be that they can see their prey better in daylight (not true, their night vision is good, and besides, buffalo are huge black shapes against a light yellow grassland; even we can see buffalo most nights).

Or, they can navigate the water better, with less crocodile activity (it's unlikely, although I'm sure that crocodiles may be more of a threat in the dark, but large tracts of this land aren't underwater, especially in the summer; the actual waterlogged area of Duba may be 20 percent at best, not enough to change their entire hunting pattern).

I can't think of other advantages of daylight over night for them.

Let's say the disadvantages at night forcing them to hunt during the day may be increased hyena activity at night (we haven't seen this, and have seen hyena takeovers during the day, so it isn't a determining factor). Increased competition from male lion takeover at night? The cooler nights slightly dissolve the daytime disadvantage to male lions. They are slightly less well adapted to heat because of their heavy and dark manes which, in some cases, increase their body temperature. But these male lions take every first kill the lionesses make even during the day.

Might there be increased nomadic male action at night to cause the switch to daylight hunting? (There are no nomadic males.)

So, all we can come up with is the Oxford theorist's third factor: that they do it because for some random reason they have learned to and it works. It really is a mystery. It certainly is not (as some scientist seriously suggested to me) that the lions are so hot during the day that they get uncomfortable sleeping and so get up (presumably out of boredom) and go hunting (as an insomniac raids the fridge for a midnight snack). These lions sleep very well, no matter what the temperature is.

I believe that the lions here hunt through the heat of the day because that is the pattern they have become accustomed to, against all odds. I wonder when they might find a new pattern.

Often the Skimmer pride poaches into the Tsaro territory, swimming in at night and snatching a buffalo before the Tsaro females are aware of their presence.

Ironically, the young males of the Skimmer pride may be the best hope for the future of the Tsaro pride, even though they are enemies right now.

Skimmer females, not quite as heavyset as Tsaro's females, still manage to live off buffalo and are skillful buffalo killers.

Following Pages: *When the Skimmer pride goes on the march across the marshy wetland, their pale bodies reflect the moonlight like pearls in the darkness.*

Pantry Pride: Stalkers in the Grass

A SOCIETY IS A GROUP OF UNEQUAL BEINGS
ORGANIZED TO MEET COMMON NEEDS.
Robert Ardrey

THIS GROUP, ALSO NAMED by the lodge guides, not by us, once numbered eight lionesses strong. One of them was quite fierce, possibly an older lion from the hunting days. Every time we saw her she made a point of walking straight up to the car and bursting into an angry charge. Once when we had a friend, Martha, with us, this lioness burst into a charge very close to Martha's side of the vehicle with a roar so loud that it made the vehicle's sides rattle (and I am sure a few other body parts within the vehicle, too). That same lioness once wandered into the camp and attacked someone walking at night.

One day the fierce lioness took on a buffalo that was just too much of a match for her, and she was sliced open across her stomach and condemned to a week of suffering before finally succumbing to death. One by one the rest of the pride fell to the buffalo or just disappeared. This pride has probably seen its natural lifespan, comparable to that of a lion, and today there are just three left. But for now they still roar their defiance to destiny from behind our camp almost every night, and occasionally swim across the river to find the buffalo.

These lions are "The Stalkers" though, and their method of hunting buffalo, probably because of their smaller group size and maybe because of the toll the buffalo have taken on them, is to sneak in around the buffalo, hiding in the grass and finally exploding in surprise, spooking the herd and then, with darting eyes, finding their opportunity.

During this project Brad, our assistant and second cameraman, was filming an attack when one of the Pantry pride lionesses caught a buffalo by the throat. She hung on during almost an hour of combat while the others alternately jumped on the buffalo's back and dodged the relentless attacks of a handful of supporting bulls that were determined to save their companion. One male put his head down and charged, hitting the lioness fully broadside over and over in an attack that would have killed most lions, and definitely would have at least made them release their stranglehold. But she took the hits and tensed her body against each successive blow. Eventually, without releasing her hold, she balled herself up under her victim's neck, and the attacking bull was hitting his companion as its head offered some shelter to the Pantry female. Finally she collapsed the buffalo, and as its calls faded away the rescuers left and the lioness moved off to the shade. It was the last we saw of her—a gallant battle, won but ultimately lost.

One member of this pride is a young male of about three years now. When he leaves, there will be only two left, and I would say that a pride of two is no longer a pride of lions. It will be over for the Pantry pride.

The Pantry pride was once as large in number but not in physical size as the Tsaro pride. But slowly buffalo attacks that went wrong whittled their numbers down to just three.

The young male of the Pantry pride will be chased mercilessly by the buffalo as well as other lions around Duba until he forges a coalition with another young male and makes a pride his own.

Pantry pride are now stalkers, mainly because they lack the numbers for a frontal attack and are easily chased off by the buffalo if detected.

Buffalo: Ancient Warriors of the Savanna

A CLICHÉ ACCEPTED TOO OFTEN, EVEN BY ETHNOLOGY, IS THE PROPOSITION THAT THE MORE DANGEROUS THE ANIMAL, THE MORE HARMLESSLY WILL HE RITUALIZE AND CONTAIN HIS AGGRESSION.
Robert Ardrey

A THOUSAND THUNDERING ANIMALS crossed the river one day in a splash of churning hooves and bellows as lost calves tried to keep in contact with their mothers and kinship groups moaned their contact in the stampede.

It is these kinship groups, these small family units, that keep the buffalo herd functioning just this side of chaos. In fact, the chaotic mass of bodies jostling through the water is quite organized into small families within larger clans within a larger herd. Like most people, at first we considered the buffalo to be a mass of wild beef, waiting (and advertising quite vocally) to be eaten by lions. But as we spent hours and days and months sitting with them through the heat and the cold we started to make out the families, recognizing individuals even in their uniformed dark bodies.

Then one day in the rolling green savanna of a Duba summer, we started talking about where we'd seen images like this before. We noticed their similarity to the great 19th-century battlefields we have come to know in paintings from around the time of Goya and El Greco.

The scene was laid out in front of us, a regiment of soldiers encamped in the fields readying themselves for the battle they knew they would soon be called to. One unit was walking through the others to take their turn at the water, while another group slept in a platoon against the hillock. Yet another stream of buffalo marched in platoon formation to the east. A few sentries were placed on the high ground and outskirts, solidly built individuals with sharp eyes and furrowed brows, scanning for the enemy. Yet another platoon was joining the regiment from the west, swelling the contingent substantially. In the rear, one platoon were rubbing their horns against the capassa trees like a special unit sharpening battle-dulled weapons. All waited with the wariness of many days spent at war, the hardships of the daily tension, and the pressures of conflict. Their gait was that of a tired army, with more than a few moving with the telltale head movement of a slightly quicker release of weight off an injured limb. The lions would see these signals at more than a thousand paces and home in like lethal missiles, and we have to wonder what pain threshold these old soldiers must wade through to have to reveal that weakness at all. They are not like us in this; often they can recover from what would be extreme pain in our bodies, perhaps being anesthetized as a result of some flight-or-fight response to attack.

This is what the herd is, a group of animals assaulted by nature and battling to survive each day, each battle, each moment. And yet—given that they are "just buffalo," we expected less of them. Our journals are filled with daily entries recording events of an attack that may change in mid-stride. Knowing that these are bovids and herding animals, we were surprised to see some of their gallant and bold defenses, and even more so their completely selfless attempts at rescuing fallen herd members.

There is an ancientness about the way buffalo move and look and go about life, just as giant ancestral buffalo once moved through the bush, masters of the emerging savannas and disappearing forests.

Buffalo shadows in a winter sunset visually increase the herd's size, fooling the eye and painting a picture that symbolizes the massive herd more than it reveals the details. These moments always remind me of the soul of the place hiding delicately behind a veil.

Great electrical storms cast lightning bolts down to earth all around the buffalo, which stoically graze on through the frightening explosions.

One such storm tore down over a hundred trees around us as we were parked out in the open grassland.

AN INCIDENT IN MAY 2005 was a good example of the way buffalo defend their own.

As we arrived, the lions of Tsaro pride were on the hunt in the first paling of dawn. The herd knew it was under attack and it was edgy, bunched together and stamping the ground into a blue haze.

Four lions moved in fast, running and looking for any weakness. It was a test, their technique in the absence of an identified target. They pushed the herd and made it run, fully aware that their first attempt would be fruitless, but the object was to get the herd on the run and watch. This time nothing looked easy, no limping animals and no calves left behind. The herd ran and circled, and then they spotted her. It was a female who looked quite energetic and alive, but the lions were on her fast and hard. Three lionesses attacked; one hung onto the throat in a death grip. We filmed and photographed what we expected to be a quick kill. But in the background we saw the herd turn, a switch from raised fleeing tails to a down-turned phalanx of horns. First one lioness saw the approaching wall and released her leg hold, then the second opted for a retreat, but the last waited to do as much damage as she could. By then the kinship group (which is usually about eight buffalo) had grown to clan size, and seventy buffalo were bearing down on her like a thunderstorm. She released her hold at the last minute as buffalo horns slashed the air about her head and sharp hooves dug into the ground around her dancing feet. The victim was bowled over by the cavalry charge but found her feet again quickly before the fleeing lions could turn and identify her and lock on for the second wave.

Soon she was just another bustling shape in the dust, and although they could taste blood and sense victory the next hour of chasing was in vain: the buffalo cow was saved.

Often we have seen the lions stalk through the grass and surprise a mother and newborn calf as it stands up, still covered in wet afterbirth. The lions know an easy meal even when it is less than enough to feed the whole pride, but sometimes a mother's defense is much more than they anticipate.

She saw the lions coming and stood over her baby, waiting. As the first lion ran in she didn't wait or run but charged, missing the lion and leaving the baby momentarily unprotected, which the second lion saw. The knockdown was swift but the mother swirled and attacked. The lioness went flying and the buffalo followed through, running over her baby to get to the lion. The third lion was at her heels and knocked the calf down again, and the mother swung around to its bellows like a dervish using all its charms and curses to beguile its enemies. But the calf was done for. We could see that it was over, but the loss to the mother was too great to bear. Something was raging in her brain. She turned and ran off a few paces and watched the lions around her baby, then made her fatal decision.

She raised her nose and from the shoulders lowered her head for the charge, which like that of the Light Brigade was doomed from the start. She narrowly missed skewering the first of the lions, but the other three were quick to dive off to either side. On the return run she drove her horns into the wild bean bushes with such force that greenery fluttered into the air like confetti, and quite suddenly the battle became surreal, almost a celebration for a moment, a reverent display of great bravery. But a crouching lioness snapped us all back to reality by side-stepping quickly and flinging herself on the recoil to land on the buffalo's back.

The bellow, usually a plea for help, seemed more pained that day, a call out to the gods in frustration as her head flew back and she looked up at the sky, with lions on her flanks. In a second or two her head sank down into the foliage alongside her baby. Through a small gap we could see a lioness lean in and start to suckle on milk from the still kicking buffalo, and we didn't approach any closer. What was going on in those dark bushes seemed somehow private and filming was inappropriate.

So altruistic are these buffalo that they act often to their extreme detriment, like the cow with her newborn calf. Although one is less likely to attach anthropomorphic sentiments to buffalo than to lions or elephants, they have characteristics that do show much more than what we once assumed of buffalo. We may not know what they are feeling exactly, as I have said earlier, but given that we have a knowledge of what we feel, this may not necessarily be a bad place to start. Emotions in our species are almost extensions of our logical thought, but the fact that we still have emotions is testament to the fact that we need them. It is also unlikely that this behavioral "invention," if it has survived and serves us well, is uniquely ours. We are, after all, part of the animal world. That animals also have emotions is without doubt. While the instinct to defend your young is strong, the sturdiest of imperatives must be the one to survive. The act of protecting a family member or newborn calf to the point of risking one's own life must surely have its motivation in some emotional source.

Exciting tropical storms sweep through the plains like traveling dervishes dancing their dance of destruction. The buffalo and lions are at the whim of the storms.

Grouped for comfort and companionship, the herd breaks down into kinship groups. Egrets and herons, wattled starlings and quelea, and the raucous oxpeckers are ever present.

White clay is found only in the north of the buffalo's range, and is an indication of where they have been.

Opposite: *Flashes of white feather fans are always a distant signal to us where to find the herd; the black bodies, even of a thousand buffalo, sometimes melt into the grassland and disappear.*

Traveling companions, early warning scouts, and lice larvae cleaners are tolerated by the buffalo in the form of the Yellow-billed oxpeckers.

Both bulls and cows attack bushes and rub their horns, usually as a display of force for those around.

WHAT WE KNOW ABOUT BUFFALO is that they seem to have evolved as animals of conflict, their large bosses and horns meant not only for display, but also for actual fighting. While males have more robust horns, females have them, too, and at least half of the conflicts in a herd are female-to-female issues. Male fights, of course, are usually longer and far more violent. We have seen a huge bull tossed up in the air, his forelegs and shoulders off the ground before being flipped over just by an interlocked set of horns and a head toss by his opponent. Mating and breeding sets off a sequence of jousting between bulls around the onset of winter. Births start around the beginning of the rains in summer. In between, there seems to be a general sense of peace and contentment within the kinship groups.

However, it is more the collective animal called "the herd" that is most impressive about buffalo. The herd moves with a unison common to many cellular organisms, such as the great flocks of millions of quela, small birds that turn on a wingbeat as one without bumbling into one another and falling out of the sky. Buffalo do bumble into one another, and kick up quite a fuss about it when they do. But to watch a herd move as one in the dusty sunset is mesmerizing, and a moment of collective beauty. It would be a mistake to view these animals simply as "the prey" here.

The lions came out of the grass as if suddenly throwing off invisible cloaks right at the herd's feet. The one-year-old calf wasn't even on the run when the first lioness squeezed in between her and her mother, literally knocking against the mother on the way to the calf. Dust hid the extent of the pandemonium, but buffalo bellows echoed over the hoofbeats as the herd exploded in panic. The slight wind drifted the dust off the herd more quickly than it did most days and revealed the buffalo standing there, noses up, trying to understand what had just happened. Eight lions in the grass muffled the calls of the dying calf under them, and the herd took some time to locate the source of the sounds. Then a wall of black bulls charged forward, heads down, intent on revenge or rescue, I don't know which.

The lions growled at each other as they do when the surface area of the food is smaller than the shoulders and mouths of the lionesses pushing and shoving for their share, but slowly the tone changed to "external" growls as the bulls charged closer. As the first lioness lost her nerve and ran, the rest of the herd gained strength from it and ran in behind the bulls, hundreds if not thousands of drumming hooves and snorting fury. The other seven lionesses lasted a token few moments and narrowly escaped being trampled. The calf was less fortunate. Already badly mauled, partially eaten, and now stampeded into the dust, his chances of survival were remote, but suddenly a hole opened up in the herd and he stood, dazed and not well, but on his feet.

The herd then divided, some individually chasing the lions, while most gathered around and smelled the fallen calf, sniffing and licking the smells of lions' breath and saliva and blood. What was going on in those minds? I don't know, but certainly something was. The herd's attentive licking may have simply been for the taste of the salt in the blood, but I think it is only part of the explanation. Maybe some day science will find a way to understand what goes on within the dark boundaries of animal emotions.

The lions retaliated four times, knocking the calf down over and over again, until finally the herd ran off far enough to allow the lions to snatch their kill. Something had changed; there was some signal, perhaps, that this was a lost cause, and the herd broke into a run over the rise. We sat there filled with the experience and the knowledge that we were observers but outsiders, locked out from complete understanding, and yet knowing that "feelings" of some sort were at play.

A wall of lions going into the hunt always spells excitement. There is a moment when the lions seem to have decided that they will hunt, and they will do it together, a chilling sight if you are their prey.

There is a moment when the investigation of a herd by lions turns into a hunt. It all starts with that one coordinated "agreement": a chase, a parry, a defense, a counterattack, and a kill.

Rescue missions that go wrong forfeit the attacked calf but also place the mother in danger. Quite suddenly the hunt changes pace and the lions switch mode from escaping the slashing horns of defense to a renewed attack on an adult. It can turn in an instant, a fatal instant.
Opposite: *No kill is easy, all require negotiating a path through the defensive mothers and other herd members.*

The buffalo escape into deep water sometimes, dragging the lions off the herd's heels by the sheer weight and depth of water. It's a risky tactic; in the water their own calves are sometimes trampled, lost, or drowned.

The Hunt

THE INDIVIDUAL ANIMALS WE WATCHED TESTED THEIR SURROUNDINGS, TRIED THINGS THAT THEY HAD NOT BEFORE.
THE PRESERVATION OF THIS CAPACITY TO ADAPT IS ONE OF THE CENTRAL MYSTERIES OF EVOLUTION.
Barry Lopez

EACH PRIDE, USING DIFFERENT basic techniques, hunts buffalo efficiently here. And because this relentless daylight hunting is so visible, their efficiency is misleading. Because they seem to be super-efficient, we tried to evaluate just how successful they really are. In judging this we had to consider what really constituted a hunt. Eventually we decided on the basic parameters: from when they arise from a resting period, go on a stalk, chase and hunt, and then declare it a miss (or kill) and go back to a resting period.

We had to define this because so often kills start with following (usually this would be a Skimmer pride tactic) or walking right in (a Tsaro pride technique). The buffalo herd stands or gathers together and chases or retaliates, and then finally runs. At this point the lions may attempt a kill and fail, but they carry on hunting again and again without stopping. This, by our definition, is just one hunt, even though they may try to catch three or four animals within that one piece of action. It isn't unusual lion behavior, by standards and observation throughout Africa, to run at an animal and get rebuffed a few times before either calling it off or succeeding, whether the prey is impala, buffalo, or elephant. However, by this parameter we calculated that in fact the lions at Duba have almost exactly the same success rate as lions elsewhere in Africa—somewhere around 25 percent. The difference, maybe, is that with these lions the length of the hunt is extended; they hunt and hunt and hunt, over and over, until they succeed. In other places in Africa the lions can switch prey after a few failed attempts, but not on this island. Here it is primarily one predator and one prey, day after day.

But these hunts are most interesting to watch because the setups are so intricate, so tactical, and so varied, ranging from the bold to the sneaky. I have little interest in the actual kill, because any and all predators kill, but these lions are displaying amazing tactics to outsmart the buffalo, testing their mental abilities as well as their physical ones, and that junction is where the best distinction between hunt and kill is to be found. It seems to me that the hunt is a mental effort, and the kill is mostly physical. That is why the hunt is more interesting. Not only does it take planning and knowledge of the field, confidence and knowledge of your own abilities (to cover the ground in a set time, for example), but it also takes forward thought.

Very often, as the hunt begins, we might see a lioness move off in a strange direction and we say to each other, "What is she thinking? What opportunity has she seen?" It is a chess game that requires moves to be thought out well in advance. And in the same way as a student of the game might see a chess move and silently say "Ah, the Karpov 1971 opening" we might say "Ah, the old Python Island water-crossing move." This is the hunt, and this is what they are born to: a special combination of thought and power as well as cooperation within the group. A lioness looks up and sees that a companion lioness has the gap between the palms covered, and then she moves up into a different position, one that may be more exposed and thus spook the herd toward that gap in the palms. Sometimes individuals set themselves up to be the first attacker; on other hunts they may be the herders. Each hunt is a running, shifting battle that is unpredictable once it gets going. The buffalo are smart and are desperate to survive this moment. The lions' job is to cut off that escape, to outsmart them, to force them into a position of weakness, and all the time ready themselves, like coiled springs, for the eventuality of a single buffalo breaking away and the attack it may need right then, in a split second—or the equal and opposite reaction to the sudden switchback of an angry bull.

Yes, this is the hunt!

The tactic: to split the herd and isolate the old bulls, then trip or anchor the massive bull, and finally to attack and weigh it down by sheer force.

Eyes watch from the distance, selecting the prey, calculating the risk, and weighing the reward.

The hunt begins with the intention. It doesn't end until the kill.
Following Pages: *The hunt goes through water, following the prey. It halts momentarily, until the next attempt.*

The counterattack can only slow down the relentless hunt and its inevitable finale. Today, tomorrow, or the next day, there will be a conclusion.

A wounded female identified by Silver Eye is knocked down over and over between tactical retreats. A wall of buffalo horns drives the attackers off time and again, but when the lions have tasted blood they seldom let their prey go. From that moment the end is inevitable, despite the efforts of the fallen buffalo's kinship group.

The Chase

GREAT AND TERRIBLE FLESH EATING BEASTS HAVE ALWAYS SHARED LANDSCAPE WITH HUMANS.
THEY WERE PART OF THE SPIRITUAL SYSTEMS THAT WE INVENTED FOR COPING.
David Quammen

WHEN THE TSARO PRIDE approach the herd, the encounter begins with pushing. But then something suddenly clicks. You can see it happen. One day we got to the herd and found the lions all snoozing, no more than three meters away from the sleeping buffalo. It was bizarre; they had almost become part of the herd. Now they were getting so used to each other that there was no pretext of hiding. The lions knew they would get a buffalo sometime, so they slept. I joked that, in time, the lions would sleep in among the herd, using their bodies as cushions. This really seemed extreme however; the buffalo were all resting, many fast asleep, with heads flat, and most even facing away from the lions.

Eventually, two hours later, the buffalo moved and the lions stretched and walked in. Now the buffalo retaliated, chasing the lions away, but each time they bounced back. The buffalo bunched together to defend. This went on for six hours, and then suddenly, as the buffalo started to charge and chase the lions, something changed.

The lions stood shoulder-to-shoulder and started to moan and growl. We have often seen them do this when they are defending against hyenas or against buffalo that are keeping them away from a kill they have made. However, this time the buffalo were keeping them from making a kill, and the lions had had enough. A lioness charged the buffalo. It wasn't an attempt to chase after them and catch one, it was a charge, with head up and neck back, a running posture that is a threat display (usually used within the species or against hyenas at best, and sometimes against humans). The buffalo also sensed the change, as if from a game to serious displeasure, and turned tail and ran off. Suddenly the lions were on the hunt; now they ran in with their ears forward, not back, necks extended forward. The buffalo charged off into the water. The lions simply ran in after them as if the water didn't exist. It was an exciting and dramatic moment: seven lionesses running, looking for opportunities, darting this way and that, all in the blue water. They were like sheepdogs herding these animals together, making them panic, taking stabs at their flanks to keep them bunched, and waiting for someone to fall over.

Someone did. A calf dropped down, but quickly found its feet. The instant that it took the calf to recover lost ground was too much. A lioness dived through the waves behind the herd and landed on the calf. Both disappeared into the water. As they bobbed up in the water, the cow saw her calf under the lion and splashed in to save it. Her hooves chopped at the water and all around the lioness's head. In a cascade of splashes and wild flashing hooves and horns, confusion veiled the lions and threw the herd into absolute panic. It wasn't until too late that the buffalo realized that there was yet another beige shape in among their legs, reaching for yet another of their young.

Through all this we moved behind in a chaotic zigzag, hopping as best we could from one slightly drier-looking island to the next in the swamp, pushing through whatever we saw the lions manage, assuming that our vehicle was at least up to getting through water that lions could wade through. At the deeper water, where they had to swim, we decided to abandon logic and risk crossing anyway. Strangely, we made it through the deepest and muddiest of the crossings and filmed both kills. When it is all over, and in the presence of bloodshed, we settle back and talk in hushed tones or just digest the moment.

Growling and fighting as they broke apart the second kill, the lions ended up much closer to us than we expected. The small patch of reeds we were lodged on suddenly collapsed under the weight of our wheels, and we were up to and beyond our axles in the marsh, once again. Now the lions had to accept our efforts to extract ourselves from the mud, and quickly adapted to our noisy efforts. Four hours and a few pints of blood lost to the mosquitoes later, we were free and exhausted. But for the lions this marshy habitat is their daily hunting ground; trudging through mud and forging their way in and out of cold water is what they have adapted to. It's no wonder they sleep well.

When a lone lioness stands in front of the charging bulls, it is at the risk of being trampled.

The water is difficult to move through fast. When lionesses run at their prey it takes all their strength and energy, the most likely reason for their enormous bulk.

Distractions dissolve: During the chase they are locked onto one objective, and need to be. A large bull could turn and retaliate, with fatal results.
Following Pages: *There is an intoxicating moment when a lion transforms from a standing, but alert animal to a charging, hunting, predator in action, a killer at the pinnacle of its evolutionary development, doing what it was born to.*

In aerial views the dance is much more evident, the tactics more easily seen. The Tsaro females form a wall of confrontation and move in with such confidence that the herd panics.

The Kill

I HAVE SEEN HORROR, AND YOU MAY HAVE THE RIGHT TO KILL ME
IF YOU WANT TO. BUT YOU HAVE NO RIGHT TO JUDGE ME.
Captain Kurtz/(Marlon Brando)/Joseph Conrad—Apocalypse Now

AFTER WITNESSING a few thousand kills, we have seen enough of these events to be completely in awe of the mechanics of the kill, but not often struck by the emotions of the moment. And yet, some kills still affect us both. Some are just sad.

We never know which will affect us, but it is obvious that the kills that most move us are ones where we know the individual buffalo. Because of the amount of time we spend with the herd, we do get to know them quite well. Females that defend their young, and die as result, are also more difficult to watch in their struggle. And it is interesting that each time we place human characteristics on the animals that we know, and we see them die, then we "feel" more for them. Most people who visit Africa and see our films are even more vulnerable than we are to attaching emotions to the prey and to being affected by the apparently cruel deaths these animals suffer.

While I am sure that with or without our misplaced sentiments animals do suffer and feel pain, they have their own version of it, one not exactly like ours. The buffalo will settle down to feed within minutes of a violent attack. Many will go to sleep, others ruminate on the morning's feed, but none outwardly show any signs of emotional distress. That they experience some pain and hardship is certain, though. They live a life daily under threat and under attack, and I imagine that, like members of our own species who perform amazing feats under stress and make stunning resistance to difficulties, they too can endure unimaginable hardships.

Suffocation is the preferred method of killing a large animal. Usually this takes the form of a throat-hold that cuts off the breathing apparatus. On a big bull, a lion may close off the nose and mouth by covering it with its own mouth. But on smaller animals, the babies and calves for example, there is little need to actually kill the animal. Lions only kill to eat, so if the animal can't get away, they eat. Only in areas with high hyena densities is it necessary to stifle the calls of the prey.

This means that quite often babies are eaten alive. It isn't for the fainthearted, but that is the way it is, and that is why Beverly and I treat what we do with such reverence. When a baby buffalo keeps calling out while eight lionesses fight over which piece of its meat they want to swallow down in a rush, to fill as much of their bellies as possible before the male comes walking over to take their food, one questions oneself, the apparent immortality of our species, God, and all the possible forces that could create such suffering.

The most powerful conclusion one is left with has to be an exuberance for life while it lasts.

By the same token, though, I won't tell you that the prey goes into shock and feels nothing. There is a burden that comes with our exposure to this, to witnessing it, and I won't make it easier. In *Ultimate Enemies,* our film about the way lions hunt elephants, we say that although there is no doubt that the kill is violent, there is no malice involved, no intentional cruelty. Perhaps as cold a comfort as it is, that is what makes what lions do so different from what man does.

A calf stands out, a single mournful call: an easy kill perhaps, but after a full day's grueling hunt and fight through the swamp, it is a merciful reward for the frustrated and exhausted lions.
FOLLOWING PAGES: *The hunt concludes with a masterful kill: a throat strangulation grip.*

THE DAY STARTED OFF like any other in winter. The inky blue of the night gave way to a cold dawn while we drank steamy cups of rooibos tea and waited for the lions to come out of the huddle they were in. We'd found them at the change from night to dull blue, the same shade as my fingertips. The sunrise was over fast, and the first real light fell on sleeping lions, still yawning steamy breath as they stretched. Eight lionesses and the two remaining cubs of Tsaro pride lay peacefully about 200 paces from the 1,200 buffalo. Just another dawn at Duba.

The buffalo dribbled off slowly from the front of the herd, and by the time one lioness looked up, the herd in front of them had diminished to a few hundred. Ahead, the first of the buffalo were already wading through icy water as if their legs had no feeling. The lionesses greeted each other as they had most other days, and a few played chase games to warm up. There was nothing unusual about the day.

They then walked toward the herd, and one or two of the lions dropped down into a stalk mode. This was unusual for Tsaro females, but the females behind them didn't bother much with concealment, so again it was a normal morning, so far.

The herd was moving right to left across the path of the lions now; most of them were in the water or across the small channel that fed a pool up ahead. I smiled. The pool was called Lion Pan. I looked at the lionesses, now dotted around next to us, all looking and learning from the herd, each lioness looking at different parts of the herd. "This is different today," I said. "Yes, they're so—intense," was Beverly's observation.

Then Beverly saw the male buffalo lying down at roughly the same time I saw the lioness to my left react, and as I quickly followed the lioness's pointer-like stance I could see that she was looking at the buffalo that Beverly had seen. A satellite image would have shown more than we could see clearly, as eight lions closed in through the grass on the one isolated buffalo. Each lioness had her head down, eyes focused on one thing ahead in time: the kill.

Too late the bull jumped up and skidded around to face a lioness in full flight, a missile launched from an ancient impulse. She hit his side and bounced, but another and then a third tackled his hocks and back. To our surprise he collapsed in a bundle before we could absorb anything but a flurry of action and our own impulses to trigger cameras. It was fast, but we had the scene on film.

We started up and moved a little closer, slowly, now in our hushed mode as witnesses to a death. But then we saw the two approaching bulls and a cow buffalo. They stepped forward tentatively, sniffing the air, flicking their tongues into their own nostrils.

We waited. The lions had flipped the bull over on his back, so easily even I was stunned by the swiftness and strength of the lions. I didn't notice through the lens that the "rescuers" had grown in number from three to a thousand! The whole herd was coming back, stepping forward with each groan from the upside-down bull.

The lions started to growl into their meal, tails flicking, ears back, all the postures we recognized from the old Savute days where hyena takeovers started like this: a slow gathering of forces from the opposition, from the enemy. The lions flattened against the grass behind their kill and behind each other, as the herd advanced one collective step at a time. The front row got to within five or six paces, and the fury of the lions echoed into the still freezing air around us all. One bold lioness leaped an explosive defense against the buffalo, and they backed away a pace or two. Flicking tongues into those noses again, they stepped back into the space they had held a moment ago. The growls intensified and the lions' tails flicked their displeasure. Another lioness jumped forward, but the herd had learned from the last time and now stepped forward to confront her. As a lioness in the back turned and dipped her head in submission, they attacked. The buffalo ran over the bare patch of grass separating this battlefield and across the psychological line dividing them, and even over their companion. One lioness waited as long as she could, but when it was so close that the slicing hooves started hammering all around her, she dived out of the way.

The herd was enraged now, chasing lions everywhere, stomping on the fallen bull, horning him, possibly, we thought, in an attempt to get him up, and spraying spit and anger across the field in answer to the growling and roaring lions. The sound closed in around us, and quite suddenly we were in their world, getting spit and blood on us as lions rushed around our vehicle chased by buffalo. Neither predators nor prey, or in this case, pursued or pursuers, took any notice of us. It was if a window had opened, and after years of looking through glass, the obstacle, transparent as it was, had disappeared, and we were in.

Riding a storm: A Tsaro lioness battles to control a buffalo cow on the run.

Although the big bull faced his enemies with strength, their weight and violence were eventually too much for him to drag his way out of. The seven-hour battle ended a long life that must have come close to ending many times before.

The bull is subdued by seven of the Tsaro lionesses, with a combined weight of about 2,600 pounds (1,200 kilograms), nearly the weight of the buffalo.

The calls of a fallen male draw his companions back in a phalanx of slashing horns and sharp hooves that will terrify even the bravest heart.
OPPOSITE: *The struggles for survival across these plains are an eternal battle: The buffalo fight for their right to live each moment, each breath, while the lions wrestle with their drive to survive another day's hunger.*

THE BUFFALO ROLLED ONTO HIS BELLY. Another bull hooked his horns in under his and threw his head back in an attack, or something else. The effect wrenched the bull onto his feet and the herd gathered around. Within seconds he was engulfed in a mass of black jostling bodies heading for the water.

The lions rallied and ran in. The herd took flight. The bull, left behind a little, was again cornered and swung around against the palms, fighting and slashing his horns, desperate for life. Again, on the first bellow, the herd turned and headed back toward the stricken bull. This time their approach was less tentative. They charged in, all horns, heavy bodies, and sharp hooves, and the lions recognized the aggression and dived for safety. This time, though, the buffalo quickly escorted the bull off into the herd while the back group defended.

Then we saw the buffalo gathered around the wounded bull. He had been on the ground for at least three minutes, and in that time the lions had ripped into his soft parts. These wounds were now bleeding, and the fresh blood attracted males, females, and calves, each licking at the blood and his injuries. He braced himself under the sea of attention, and we debated whether it was the salty taste of blood that so attracted the other buffalo, or something else. One idea is that it is a compassionate attentiveness; another is that the natural antiseptic in animal saliva could heal the injuries. I think it is the precious allure of salt. But the buffalo were actually quite rough and competitive in getting to lick and smell the blood. Once or twice the bull had to chase them away because he was being bumped off his feet. Then a buffalo jumped up on his back and started to mate the bull, or mock mate, because this was a female buffalo! She was quickly shaken off only to be replaced by another buffalo, this time a male. And so it went on; while the injured bull was within the herd he was mounted and licked and bumped. When he was nudged to the outskirts, the lions were waiting.

Around midday we'd been with the bull for six hours, and apart from a few sessions of about ten minutes of rest between attacks, the lions had been alert and on the advance. They pulled the male to his knees a dozen times, only to be rebuffed by the herd. They moved across the water and into the open grassland, following, attacking, diving away, dodging buffalo charges, all the while circling around to the bull.

Finally, when we thought this stalemate would never end, a single bull stepped out of the herd. I saw him coming, head held higher than the others. As I started filming again, he walked up to the male, then put his head down and attacked the injured bull ferociously, hitting him over and over on the side and locking horns with him, finally flipping him as easily as I might lift a young child. The bull crumbled under the attack and rolled over. The attacking bull, with a shine of blood on his horns, stepped back and gave him one more head butt on the ground. Then he turned his back on the injured bull, and left.

The herd followed, and within moments the Tsaro pride walked over and killed the male under the gaze of the last three buffalo, who finally turned and walked off. It had been a seven-hour battle, and the lions were relentless. The buffalo put up a good defense that could have sent any of the lions off with their own blood covering their bodies, not just the injured bull's. In battle one would salute a noble enemy. But then they don't do this for glory. There is no Pro Patria or Purple Heart medal or ticker tape parade for battles well fought, nor do the buffalo go off to plan their counterattack one day. The lions do it to eat, and the buffalo do it to stay alive: relentless enemies.

Up against the wall with nowhere to go, these lions have nerves of steel.

WHY THE MYSTERY BULL stepped out and put an end to it all, we will never know. My best guess may change, but today I think he had just had enough. He was irritated by the fuss and the handicap of being prevented from continuing with his day. It could have been a near-ranking bull who took advantage of the incapacitated senior to oust him from his position, or a mercy killing attempt (which may be taking it too far). It could be that he attacked simply because the bull now smelled of lions, or even more simply was behaving weirdly, and that was enough for him to want to be rid of it. Lastly, he could have been unsettled by the way this injured bull was attracting the lions. Certainly his gesture was final. The herd moved away after that and not before. And the window closed on us again. Sure, we could see through it, but as if coming out of a dream we were back in reality. In the distance the hum of a light aircraft delivering freight or passengers to some lodge—Jao or Mombo perhaps—a cool wind on one side of my body, and a burning sun on my leg, all conspired to swing that glass across my mind and leave me desperate to stay, but unable to. The moment of intensity was over.

At least we could still see inside, but now we were analyzing and talking about the number of attacks, the motivation of the mystery bull, and suddenly a wave of sheer exhaustion washed over us both. Some of it was just the sadness of someone else's suffering. Much of what we were feeling was just the release of our concentration on getting the right focus and exposure, and in my case the variety of angles and focal lengths needed to create a moment in time that would be as unforgettable to others as it was for us. Mostly it was the exhaustion you feel when you have been intellectually challenged, made to confront your demons, and forced to think about the horror of death, your own, perhaps.

The bull had turned to face his demons time and time again, each time with a compassionless glassy eye. At every turn, everywhere he looked, the lions were there like some midnight nightmare, and each time he could have given up. He had given up, I think, at times, but some greater force within him picked him up and made him fight back. It was for nothing, of course. And what about us?

It is dark outside as I write this. A hyena stalks around camp. I go outside and look at his tracks over mine. They are small enough, but still a haunting reminder that this is sometimes, no, always, a dispassionate place, dispassionate as to whether we are here or not. Africa will survive beyond us. We may drag down some of its other inhabitants with us, like desperate swimmers in a group overwhelmed by the waves, but this place and the spirit of it will outlive us. How will we turn and face our end, I wonder? Will we judge ourselves as we do the bull, as is natural for us humans to do? No one else will. But if we do, just how will we stack up?

The kill is the pinnacle of our journey here. It is the most intense moment, a transition from the magical and athletic to a more metaphysical realization or appreciation, from life and action to death or the unknown. All transitions, especially those into the unknown and unsafe, are exciting and invigorating. So should this one be, whether it is a buffalo's, a lion's or our own.

If we learn nothing from such an intense moment, then it is all for nothing. Only the very insensitive, those same flawed people who derive pleasure from killing a lion (for fun or "recreation") can enjoy watching the kill. All reasons other than to learn and experience some inner journey from the kill are just window dressing, avoiding the most important moment one can be exposed to, and skipping out of those lessons—a waste of time.

The day after the bull kill we didn't follow the Tsaro pride or try to find Skimmer. We went to the river and sat there for a while.

Only one lioness waited until the very last before the wall of buffalo descended to overwhelm and collect their fallen companion.

At first contact, the sheer weight of the Tsaro females is often enough to collapse a cow buffalo.

A battle of the Titans, where a large buffalo cow tries to drag her assailant off in the water, ends in defeat or discretionary retreat when the water and threat of crocodiles in the deep water overwhelm her.

In the wet season, when calves are born, lions use a different tactic. They follow the herd and watch for single mothers left behind, the signal of the birth of a calf. The easy prey are simply picked off as they are born.

When a calf is attacked it is usually the immediate kinship group that returns to rescue it. These family groups of about 20 related buffalo are formidable and aggressive, with a violent drive to rescue their fallen kin. Sometimes a lion's only defense is retreat.

ALTHOUGH WE MAY HAVE NEEDED a break from the never-ending drama, the lions did not. The bull was turned into orange dust under the feet of squabbling vultures. The nine lionesses, two males, and two now-bloated cubs had eaten every scrap and were hunting again.

The dawn was filled with the scent of Duba at its best—anticipation.

A rare interaction with hyenas started the day. First they ate off part of our winch-cable housing. Toying with three lionesses, they badgered them to the point of complete frustration, forcing the lions to get up and walk away from their intended meal: the advancing buffalo. The Tsaro females, vulnerable as just a partial pride, waded out across the water away from trouble.

Suddenly the hyenas turned their attention to the buffalo, and in a bizarre turnaround the lionesses found themselves ahead of the stampeding herd. They were ideally positioned and killed two very young calves in a dramatic spray of water that hid the calves and confused the rescue attempts of their desperate mothers. However, in what was for us a strange echo of our work at Savute, we sensed tension coming through the treeline. Shortly afterward the first drooling hyena came leaping ahead of himself in his eagerness to share in the spoils. Soon there were 13 enraged hyenas bouncing through the water, racing toward the lions. Within seconds it was over, a simple strategy of psychological aggression that overwhelmed the lions and shattered their confidence. It was one of the few hyena takeovers we'd seen at Duba. It was vicious and it was fast, almost surgical.

A kill is not always where it ends. Even after hours of pushing through water, chasing down a herd, dodging the slashing horns and thundering bodies, the lions lose their kills just about as often as they keep them. The hyenas are just one of the obstacles. The male lions are another. They could actually be considered almost parasitic on the females, if it weren't for their vigilant patrolling and protecting role.

Joseph Conrad wrote in *Nostromo* that "Action is consolatory. It is the enemy of thought and the friend of flattering illusions." Across the savannas and marshes of Duba, this "action" goes on every day, but far from being a flattering illusion, it serves to stimulate much thought in our minds.

Demonic beasts from a lion's nightmare are enough to strike fear in even the bravest.

Small but effective, the hyenas swoop in and take the lions' kills. Their intimidating, aggressive approach usually sends the lions off running.

It is strange that 13 hyenas are more intimidating to these lions than a thousand buffalo. Some kills just cannot be defended. The day belongs to the hyenas.

She is beautiful in her symbolic makeup, horrific as its meaning is. Someone has died to feed her and her cubs.
FOLLOWING PAGES: *The ghosts of buffalo are scattered across Duba Plains. The floodwaters may wash the bones apart, but something still remains.*

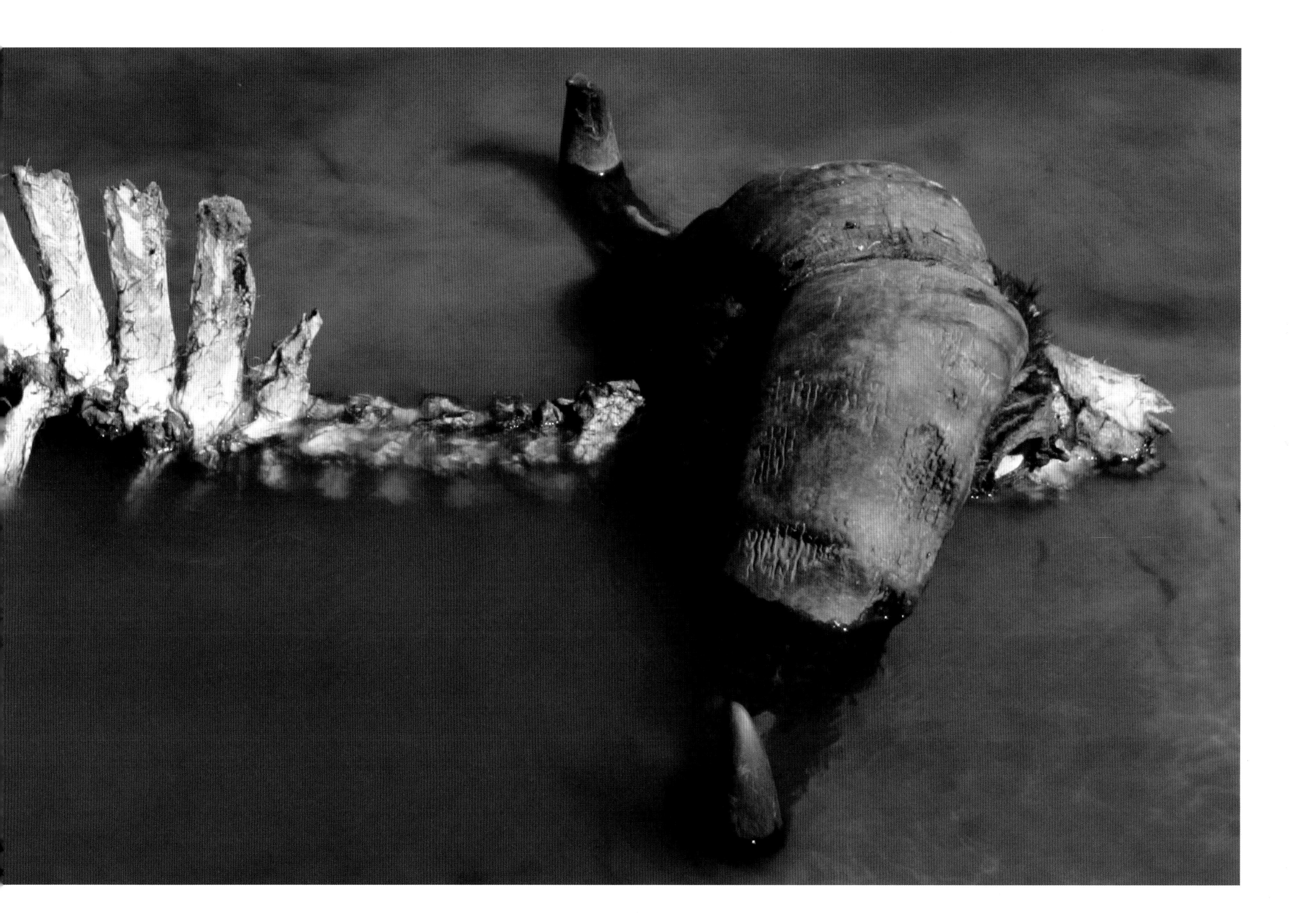

And in the End

PRESERVING LIFE SHOULD BE THE NATURAL WAY OF
COMMERCE NOT THE EXCEPTION.
Paul Hawken

THIS IS AN INDIVIDUAL JOURNEY, for you, the buffalo, the lions, and us. We will all learn differently from it. The subject is singular and obvious. It will chase us and make us confront it, but perhaps each one's time scale is different.

On the surface, the lions have learned so much. They stalk and chase and then kill according to a new set of lion behaviors, different from what we know from other parts of Africa, even Botswana. Lions learn daily. They never turn away.

Similarly, the buffalo have learned how to survive on this island, and it is a mystery why they don't just leave. In other parts of Africa where we have worked, after a lion attack, the buffalo "bounce" to the farthest corner of their range, getting as far away as possible from the lions. Here, that would send them straight from Tsaro to Pantry to Skimmer, and they would lose three buffalo a day. So instead, after an attack the herd now just settles down nearby, often to sleep. Maybe they know it is safest where the lions have just killed, maybe it is the aftereffects of the stress they go through, but often enough the buffalo do not charge away after a kill.

They have also learned that if they are under attack the best defense is to sleep! During an ongoing attack by lions, quite often the herd just pulls together into a bundle, and slowly individuals drop down for a nap that lasts an hour or two. During that time, the lions have no option but to sleep as well; attacking a sleeping herd of buffalo all bunched together with horns facing outwards, the young hidden within, is impossible. Lions use the movement of the herd to watch for defects and weak points within the herd; these vulnerabilities are also closely guarded in the huddle of resting bodies.

In the heat, this extended rest period in the sun often tips the lions' heat exhaustion levels over the edge, while for some reason the buffalo are fine in the sun, and get up and leave the panting lions behind. These lions, however, now force themselves to get up and follow, and the equivalent of an arms race continues.

The cats are in action, regardless of the water they have to cross, and regardless of the beasts they have to hunt.
FOLLOWING PAGES: *The lions wade through water for hours each day. This was a seven-hour day when they and we were seldom on dry land. We followed in our vehicle, adopting the philosophy that if they could do it, so could we. We were wrong.*

SOAKING UP EACH OF THESE MOMENTS like hungry students is our way of processing what can, at times, be quite hard. The life that Beverly and I have chosen is a difficult one, often on the edge, often uncomfortable, but what we most enjoy about it is that it takes us to that meeting place of past instincts and present intellect. Only right there, at that point, are we truly alive.

None of us, though, no matter where we choose to live, can escape the primitive complexities we have inherited from our simian past, and we get involved in the intricacies of our small hardships. We sometimes draw down the blinds on the richness of each consecutive breath that automatically fills our lungs. Like a diver who has run out of air underwater once, and forevermore savors each breath, we drive through these fields littered with the bones and ghosts of the fallen and fill our souls with the reality of being alive, now. If nothing else, that should be enough.

But it isn't; our heads fill up with the pressure of being creative or just plain clever, and we study Descartes or the mathematical Spinoza, Jung and Freud, in some hope that they understood better, when in fact it is all inside and perfectly understood by that part of us that has been around for a few million years.

The relationship between the lions and the buffalo of Duba shifts and sways back and forth, changing from time to time and returning to what it was. These two ancient and relentless enemies are locked into this dance, a parallel to our own troubled relationship with nature. As a species we fight it into apparent submission daily, but it bounces back at us. We have developed our understanding enough, however, to realize that it is the dance that is important, not who wins or loses.

And far from being disillusioned about the future of the dance, I have to confess to a great optimism. I feel the slow wind of change, just a breeze against the skin for now, but there nonetheless. Without hope, the future will be no better than the past, and we can't live with that.

The buffalo herd has done well despite the heavy predation. Numbering well over a thousand now, it grows at a rate of about 5 percent per year.

FOLLOWING PAGES: *The Duba lions' destinies are locked into those of the buffalo: without their constant presence and supply the lions would not survive here—an island with few other animals.*

Her enormous bulk weighs heavily on a Tsaro lioness leaping over a stream.
Opposite: *Tails paint the air with water brushstrokes as the Tsaro females fight for a piece of meat. Lions are unique in that they cooperate in the hunt but compete for the food.*

The flood is at its lowest in summer; the rains release a long chain of events that bring floodwater to the Okavango only in winter, six months later. When the buffalo are not at the rivers on either end of Duba, rainwater pools attract them in the wet season between floods.

FOLLOWING PAGES: *Against a tapestry of reeds and papyrus, a Tsaro female searches the distance for signs of buffalo returning to her territory.*

The glow of evening brings relief not only from the heat but also from the relentless hunting of lions, as at least 90 percent of the lion predation happens in daylight hours.

MANY PEOPLE WORK TOGETHER to get a project like this done.

We asked the Okavango Community Trust, the leaseholders of the Duba concession, for permission, and received it within a week. We need to thank them for their foresight in seeing the potential of this project. Numerous people have come to our aid, when after hours (or in one case days) we reluctantly walked for help, or later used the modern invention (a radio) when we were stuck up to our axles in mud, bogs, water, peat forests, and just about any place you can imagine outside of a reasonable radius of anything solid to winch against—usually beyond the reach of wood of any kind, and mostly in an ever-softening surface. This is not an easy place to work in.

We would like to thank Colin Bell, "James" 007 (Kebalibile) Piseru, and Chief and Tumpane Lenchwe Maolosi at Duba camp, who are our guides, managers, friends, and just good people. Paul De Thierry at Duba has come to our aid often enough and has been a good neighbor during these years.

Also we thank Jan Broekhuis, Rapelang Mojaphoko, Director of Wildlife and National Parks, and, collectively, Botswana's Department of Wildlife.

We thank President Festus Mogae, our own Tau e Tona (Great Lion), a great appreciator of the wildness of nature here, and custodian of that cause; the Office of the President, which has become a hub for us in many ways; and indeed the people of Botswana, who against overwhelming odds struggle to forge ahead and refuse to succumb to the gale force of corruption that howls through so many places in Africa. I congratulate every attempt to stand upright against this scourge, which in other places is accepted as "just the way things get done." Not here. Thank you.

There are actually few examples of wild places, national parks, and game reserves that work quite as well as those in Botswana. Sometimes the public facilities may not be perfect and the roads are terrible. But Botswana is 100 meters ahead in the 100-meter race to understanding that wild places sometimes need to be left alone. Ecotourism is the way ahead, but not in the heavy-handed, over-managed styles of some other countries, and local communities best serve the environment when they have a meaningful, long-term stake in it. Most of that amounts to giving stakeholders enough space to run their own shows within a set of guidelines. Sometimes that may appear like no management, but the results are clear: The wildlife and the tourism experiences in Botswana are unequaled in the world.

Brad Bestelink and Andy Crawford need special mention here as our longtime assistants and supporters, production crew, and now post-production team, Brad having filmed some of the sequences in the associated film, *Relentless Enemies: Lions and Buffalo.* I want to thank Keith Joubert, my brother, for years of influence and, in particular, inspired discussions infused with creative challenge—proof that art and science, far from being opposite disciplines are instead, if used appropriately, completely mutually compatible. At our best, we use one to feed the other in a nurturing way that makes great art and creative science.

Tanya Ritter, Keith Vincent, Mike Wassung, and Malcom McCollogh have all been instrumental in making our lives easier out in the bush, as have all of the Okavango Wilderness Safaris staff in general, and we thank them collectively. Lorna Gibson runs our lives better than we could, and Sarah Rachmann has designed the book that we wanted with dexterity and creativity, with Robbie Frey of Resolution Colour for reproduction work.

Once again we must thank the National Geographic Society and National Geographic Channels for their support, as well as Chris Johns, Kevin Mulroy, and the ladies: Becky Lescaze, Leah Bendavid-Val, and Marianne Koszorus of the Book Division at National Geographic.

To Ian Khama, friend and associate, we thank you for more things than there is space for here. You lead, inspire, care, and offer a paternal wisdom with a friendly touch to the nation and beyond.

In an ancient ritual between buffalo and lion, the battle rages relentlessly, at times in favor of one and then shifting in favor of the other in a dramatic duel of equals.

Relentless Enemies

Dereck and Beverly Joubert

Published by the National Geographic Society
John M. Fahey, Jr., President and Chief Executive Officer
Gilbert M. Grosvenor, Chairman of the Board
Nina D. Hoffman, Executive Vice President;
President, Books and School Publishing

PREPARED BY THE BOOK DIVISION
Kevin Mulroy, Senior Vice President and Publisher
Leah Bendavid-Val, Director of Photography Publishing and Illustrations
Marianne R. Koszorus, Design Director

Barbara Brownell Grogan, Executive Editor
Elizabeth Newhouse, Director of Travel Publishing
Carl Mehler, Director of Maps

STAFF FOR THIS BOOK
Rebecca Lescaze, Editor
Michael B. Horenstein, Production Project Manager
Cameron Zotter, Design Assistant

Rebecca Hinds, Managing Editor
Gary Colbert, Production Director

Pre-press production by Resolution Colour Pty (Ltd), Cape Town, South Africa
Layout by Sarah Rachmann, Enlightenment

MANUFACTURING AND QUALITY MANAGEMENT
Christopher A. Liedel, Chief Financial Officer
Phillip L. Schlosser, Vice President
John T. Dunn, Technical Director

Wildlife Films Botswana Pty (Ltd)
www.wildlifeconservationfilms.com

ISBN 10: 1-4262-0004-8
ISBN 13: 978-1-4262-0004-5

Books by the Jouberts:

Hunting with the Moon
Whispers an Elephant's Tale
The Africa Diaries
Elephant in the Kitchen

Films:

The Stolen River
Trial of the Elephants
Journey to the Forgotten River
Patterns in the Grass
Reflections on Elephants
Lions of Darkness
Wildlife Warriors
Whispers an Elephant's Tale
Hunting Hounds of Arabia
Eternal Enemies
Ultimate Enemies
Relentless Enemies
Eye of the Leopard

Founded in 1888, the National Geographic Society is one of the largest nonprofit scientific and educational organizations in the world. It reaches more than 285 million people worldwide each month through its official journal, NATIONAL GEOGRAPHIC, and its four other magazines; the National Geographic Channel; television documentaries; radio programs; films; books; videos and DVDs; maps; and interactive media. National Geographic has funded more than 8,000 scientific research projects and supports an education program combating gegraphic illiteracy.

For more information, please call
1-800-NGS LINE (647-5463)
or write to the following address:

National Geographic Society
1145 17th Street N.W.
Washington, D.C. 20036-4688 U.S.A.

Log on to nationalgeographic.com;
AOL Keyword: NatGeo.

For information about special discounts for bulk purchases, please contact National Geographic Books Special Sales:
ngspecsales@ngs.org

Library of Congress Cataloging-in-Publication Data
Joubert, Beverly.
Relentless enemies : lions and buffalo / photographed by Beverly Joubert ; and written by Dereck Joubert.
p. cm
ISBN 1-4262-0004-8
1. Lions—Botswana—Okavango Delta—Pictorial works.
2. African buffalo—Botswana—Okavango Delta—Pictorial works. I. Joubert, Dereck. II. National Geographic Society (U.S.) III. Title.

QL737.C23J68 2006
599.757096883—dc22

2006046160